Jerzy Sobczak

Ludmil Drenchev

Metal Based Functionally Graded Materials

Engineering and Modeling

Contents

FOREWORD

The concept of functionally graded materials (FGMs) introduced a class of highly engineered structures tailored to specific properties, resulting of compositional changes in used materials. The necessity to bring into practice new materials appears crucial with, for instance, space vehicles: on the surface side the skin plates should have very good heat-resistance, on the inside, however, – high mechanical qualities (e.g. toughness) were needed. The problem was successfully solved in Japan in the mid of 1980s by manufacturing specific composite: metallic matrix and ceramic particles with graded distribution of these particles. That solution is close with ingenious structural systems in some plants, e.g. bamboo. After 20 years of intensive research and practical applications, the field of FGMs is still in development and a precise definition of that new class of materials is till now not accepted. Modeling of FGMs is recognized as indispensable step in designing at the microstructural level to meet specific requirements of an intended application. Many production technologies were proved to be useful for practical adoption.

The authors of the book discuss two important topics concerning activities within the field of FGMs. One of them is dealing with practical methods used in graded materials technology to control the composition and microstructure. A considerable number of problems belong to the field of FGMs manufacturing and a rich source of motivation stimulates in some cases application of new specific fabrication techniques. The book brings condensed but valuable information about main solidification and sintering techniques, which can be used for appropriate manufacturing solutions.

Second part of the book gives theoretical approach to the mathematical modeling and design of FGMs: appendixes A, B, C and D are presenting the authors' novel research results.

The authors deserve congratulations for a very compressed but useful information source.

Stefan Wojciechowski, *D.h.c., Ph.D., D.Sc.*
Warsaw University of Technology
Chairman of Composite Section
Polish Academy of Sciences
Poland

PREFACE

Functionally graded materials, their characterization, properties and production methods are a new rapidly developing field of materials science. From the most informative and widely acceptable opinion, functionally graded materials are characterized by gradual space changes in their composition, structure and, as a result, in their properties. Usually, they are composites in the common sense, but graded structures can be obtained also in traditional, monolithic materials on the basis of a variety of microstructures formed during some kind of material processing. These materials do not contain well distinguished boundaries or interfaces between their different regions as in the case of conventional composite materials. Because of this, such materials posses good chances of reducing mechanical and thermal stress concentration in many structural elements, which can be developed for specific applications. The structure is not simply inhomogeneous, but this inheterogeneity is usually in one direction, typical for the entire volume of a material. The development of instruments for micro- and macrostructure design in functionally graded materials is a challenge for modern industry. On this path, mathematical modeling and numerical simulation are extremely helpful tools for design and investigation of functionally graded materials, which, in fact, are typical representatives of knowledge-based multiphase materials.

The aim of this book is to provide a comprehensive overview of the basic production techniques for manufacturing functionally graded materials, with attention paid to the methods for quantitative estimation of the main structure parameters and properties of these materials. A concise description of experimental methods and type analysis of some specific structures obtained are presented. In order to provide an easily-readable text, general mathematical models and specific tools for management of graded structures in metal matrix composites are presented separately in appendixes. The movement of particles during gravity and centrifugal casting is widely discussed in Appendix A and Appendix B, respectively. Basic equations, which describe solid particle movement in the case of Lorenz force application in graded structure production techniques, are given in Appendix C. A comprehensive model of physical phenomena in gasar technology can be found in Appendix D.

It is our hope that this book provides valuable information for all colleagues who interested in the field of functionally graded structures and materials, and who need a compact informative overview of recent experimental and theoretical activity in this area.

This book has been published as one of the results of Commissioned Research Project PBZ/KBN/114/T08/2004 "Innovative materials and processes in foundry industry" sponsored by the Polish Ministry of Science and Education (Task II.2.4. The study of technology of graded metal-ceramic products by advanced casting techniques, including use of external pressure).

We would like to thank Bentham Science Publishers, particularly Director Mahmood Alam and Manager Bushra Siddiqui for their very kind support and efforts.

Jerzy J. Sobczak
Professor
Foundry Research Institute
Krakow, Poland

Ludmil B. Drenchev
Associate Professor
Institute of Metal Science
Sofia, Bulgaria

CHAPTER I: MAIN PRODUCTION TECHNOLOGIES

Ludmil B. Drenchev[1], Jerzy Sobczak[2]

[1]*Institute of Metal Science, 67, Shipchenski Prohod Street, 1574 Sofia, Bulgaria,* [2]*Foundry Research Institute, 73, Zakopianska St., 30-418 Krakow,Poland*

Address correspondence to: Ludmil B. Drenchev; [1]*Institute of Metal Science, 67, Shipchenski Prohod Street, 1574 Sofia, Bulgaria; Telephone: +359 2 46 26 223, Fax: +359 2 46 26 300, Email: ljudmil.d@ims.bas.bg*

Abstract: This chapter deals with the concept of graded material and main technologies for their production. The technologies for metallic graded materials are divided into two general groups. A sequence of eighth basic production methods is discussed in more details.

I.1 Materials of Graded Structure

The idea of Functionally Graded Materials (FGMs) was substantially advanced in the early 1980's in Japan, where this new material concept was proposed to increase adhesion and minimize the thermal stresses in metallic-ceramic composites developed for reusable rocket engines [1]. Meanwhile, FGMs concepts have triggered world-wide research activity and are applied to metals, ceramics and organic composites to generate improved components with superior physical properties [2]. Depending on the application and the specific loading conditions, varying approaches can be followed to generate the structure gradients. Consequently, coatings have been deposited by different techniques involving physical vapor deposition (PVD), chemical vapor deposition (CVD), plasma spraying, arc spraying, pulsed laser deposition (PLD) or sol–gel techniques.

Today, production of graded structures can be considered as the next step in composite materials development. Functionally graded materials are a relatively new class of engineered materials in which the composition and/or microstructure varies in one specific direction. They are made by a continuous change in composition and do not possess a specific interface. Therefore, it is generally assumed that such a structure should better resist thermal and mechanical cycling. The application of this concept to metal matrix composites (MMCs) leads to the development of materials/components designed with the purpose of being selectively reinforced only in regions requiring increased modulus, strength and/or wear resistance. The graded structure means graded properties, and the obtaining of such structures extends to an essential degree the industrial applications of the materials considered, and especially of MMCs.

The aims of this review are to systematize the production methods and to generalize the principles for obtaining graded structure in MMCs. Special attention will be paid to centrifugal casting as the most productive and controllable method, to the application of electromagnetic field for composite structure government and to the obtaining of a graded porous structure in gas-reinforced metal matrix composites (gasars). Some details regarding settling in gravity and squeeze casting will also be discussed. Mathematical models of the most essential physical phenomena in the five above mentioned casting methods will be presented in applications and it will be place special emphasis will be placed upon physical and technological instruments for structure control.

Since nonmetallic FGMs are not the focus of our consideration, only certain aspects of their production technologies will be mentioned.

All methods for the production of MMCs can be used in a more or less effective way for obtaining of inhomogeneous structure, but not all of them have the potential for flexible impact on the structure obtained. It seems that centrifugal casting and electromagnetic field utilization are the two methods for effective control of particle distribution in MMCs. Squeeze casting is the most popular casting technology for infiltration of molten metals into graded preforms. Special methods are also being developed for surface deposition of structured coatings.

Generally, there are two methods of graded structure management in MMCs production. The first one consists of preliminary preparation and/or arrangement of composite's components, and after this, by means of a certain melt processing technique, obtaining a final product. Typical examples are: melt infiltration in a graded preform, sintering of gradiently stacked powders, and consecutive mold filling by melts of different composition. The second method is based on the application of one or more external force fields that exert different influence on the separate components in a liquid composite slurry. Such fields can be gravitational, centrifugal, magnetic, electrical (at electrophoretic deposition) and electromagnetic. Special instruments for management of porous structure in gas reinforced MMCs (gasars) are external gas pressure and solidification velocity.

I.2.   Structure Management by Gravitational Force (Gravity and Squeeze Casting)

Sedimentation and flotation are the basic phenomena in all technologies for production of particle reinforced composites. As gravity is present everywhere on earth, these phenomena cannot be avoided, but can be exploited in the preparation of graded structures. Particles of different density and mass move differently in liquid metals and alloys and by appropriate thermal control of die cooling, which means control on the magnitude and direction of solidification velocity, graded structures in MMCs can be obtained. Theoretical aspects of such processes can be found in [3] and are discussed in Appendix A.

The influence of gravity on particles in a powder mixture is used successfully in the so-called *co-sedimentation* process for preparation of specific preforms for melt infiltration. This process is one of the most promising ways to fabricate large-scale FGMs with smooth variations in composition and microstructure. A kind of W-Mo-Ti FGM with density gradient has been fabricated by a co-sedimentation method [4]. This is done by preparing the deposit body after settling of the corresponding powders layer by layer. The pure Ti layer is settled first, then the Ti-Mo graded layer and the Mo-W graded layer, and the pure W layer is settled last. Green and fine metal powders of W, Mo, Ti, Ni and Cu are used, in which Ni and Cu powder acted as sintering activators. After appropriate powders treatment a set of suspensions are prepared. Then the suspensions are poured successively into a sedimentation tube. After complete particle sedimentation, the deposit body is taken out integrally and compacted. Then the compact is sintered at 1473 K under a pressure of 30 MPa in a vacuum furnace. The structures obtained can be seen in Fig. 1 and Fig. 2.

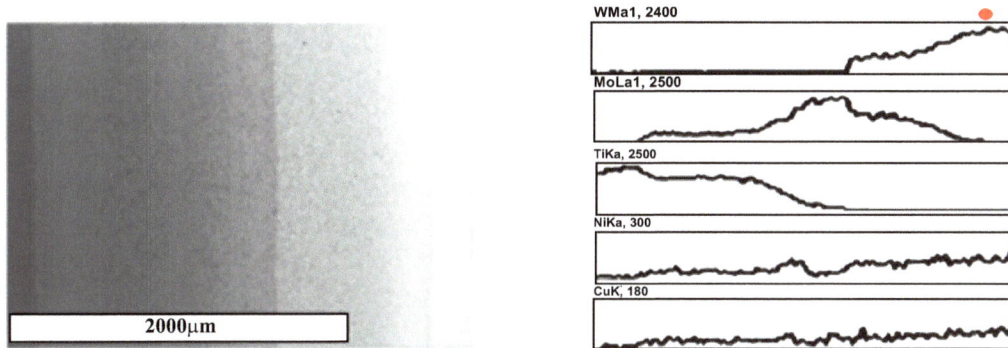

Fig. 1.   Electron image and linear distributions of elements along the cross section of W-Mo-Ti FGM (Ref. [4])

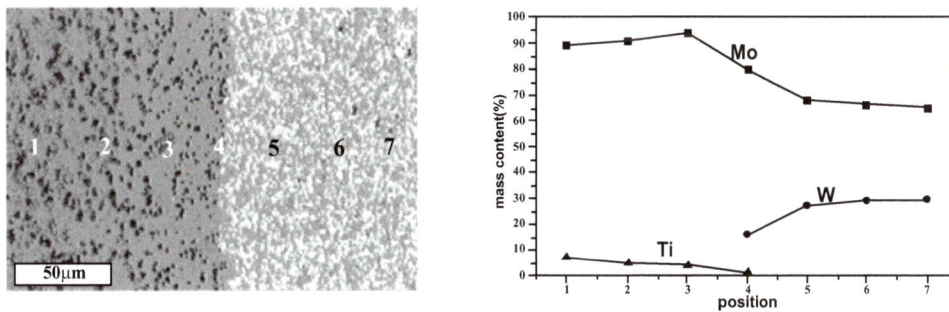

Fig. 2.   Electron image of Ti/Mo and Mo/W interface and element contents (wt.%) in a different position (Ref. [4])

I.3.   Structure Management by Centrifugal Force (Centrifugal Casting)

In centrifugal casting, besides gravity, *centrifugal* and *Coriolis forces* act on the particles, which are dispersed into the melt. In almost all cases gravity is very small with respect to the centrifugal force and can be neglected. There are specific conditions under which the Coriolis force can be also neglected. A mathematical description of centrifugal casting of particle reinforced MMCs and details of its application are given in [5,6]. Some theoretical aspects of graded structure manufacture by centrifugal casting are analyzed in Appendix B.

Functionally graded aluminum matrix composites reinforced by SiC particles are attractive materials for a broad range of engineering applications whenever a superior combination of surface and bulk mechanical properties is required. In general, these materials are developed for the production of highly wear resistant components. This kind of mechanical part often operates in the presence of aggressive environments. There are a great number of papers [5-14] devoted to different aspects of centrifugal casting of particle reinforced MMCs and its potential for production of graded structures. This method may be considered one of the most effective techniques for production of particle reinforced FGMs.

Sequeira et al. [15, 16] apply centrifugal casting for obtaining Al-Al$_3$Zr and

Al-Al$_3$Ti functionally graded materials from Al-5 wt. % Zr and Al-5 wt. % Ti commercial alloys, respectively. The alloys are heated to a temperature between solidus and liquidus temperatures, where most of the intermetallic platelets remain solid in a liquid Al-based matrix, and then poured into a rotating mold in order to obtain cylindrical samples. The intermetallic platelets are orientated predominantly perpendicular to the centrifugal force, and their concentration in the outer casting region is essentially higher than in the inner region, see Fig. 3. An increase in hardness up to three times is observed in intermetallic rich regions.

The formation of compositional gradient during manufacture of FGMs by the centrifugal method is studied in [14]. Al-Al$_2$Cu FGMs are *in-situ* fabricated using eutectic Al-33 wt. % Cu alloy. The microstructure of fabricated FGMs is analyzed on the basis of the Al-Cu phase diagram.

Functionally graded boron carbide/aluminum composites are obtained by both centrifugal casting and type casting methods. The microstructure as a function of different processing parameters and the corresponding spatial variations of hardness are characterized in [17].

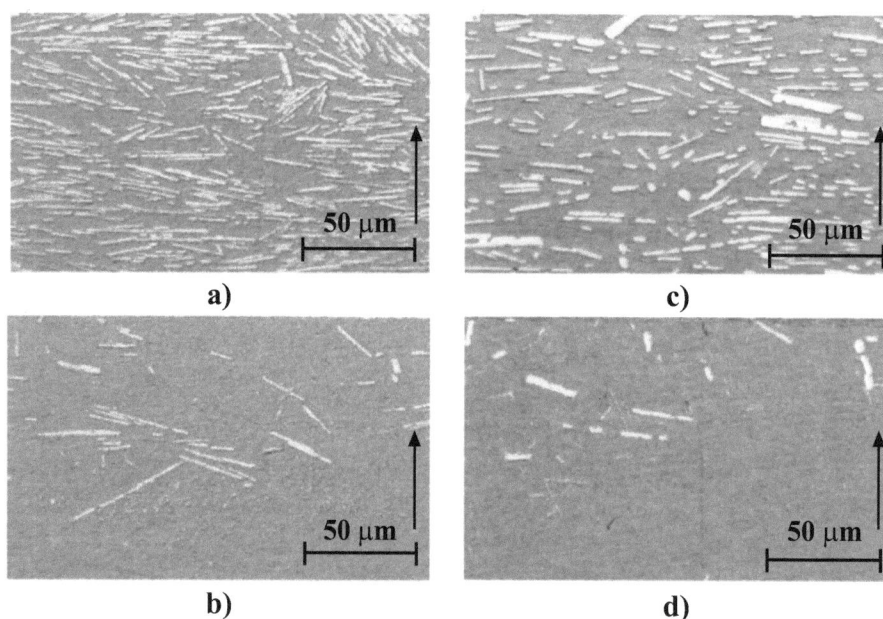

Fig. 3.   Optical micrographs of different regions, brighter areas are the intermetallic particles:
a) and b) are from Al-Al$_3$Ti FGM in outer and inner regions, respectively;
c) and d) are from Al-Al$_3$Zr FGM in outer and inner regions, respectively.
The arrows indicate the direction of applied centrifugal force (Ref. [15])

Song et al. [18] studied the microstructure and properties of graded WC$_P$ reinforced ferrous composites (WC$_P$/Fe-C) obtained by centrifugal casting. The microstructure, mechanical properties and wear resistance of the gradient composites are investigated. The results are compared with those of high speed steel. It was found that the tensile strengths of the gradient composites layers reach 460 MPa. Moreover, the hardness of the composite layers reaches HRA 81. The results of comparison between the graded composites obtained and high speed steel show that the wear resistance of the gradient composites layers are more than 20 times higher than those of the high speed steel under loads of 100 N and 200 N and sliding velocity of 60 m/s. The structure of different composite layers can be seen in Fig. 4.

a)   b)   c)

Fig. 4.   Distribution of WC_P in composite layers at 800 rpm: (a) 70 vol.% in outer layer, (b) 60 vol.% in inner layer, (c) transition layer between the gradient composites layer and particle free region of the Fe-C alloy matrix (Ref. [19])

An experimental study of the effects of the rotation velocity in centrifugal casting on the graded structure and properties of the same WC_P/Fe-C composites is described in [19]. The concentration of WC_P, hardness and impact toughness at different radii are shown in Fig. 5.

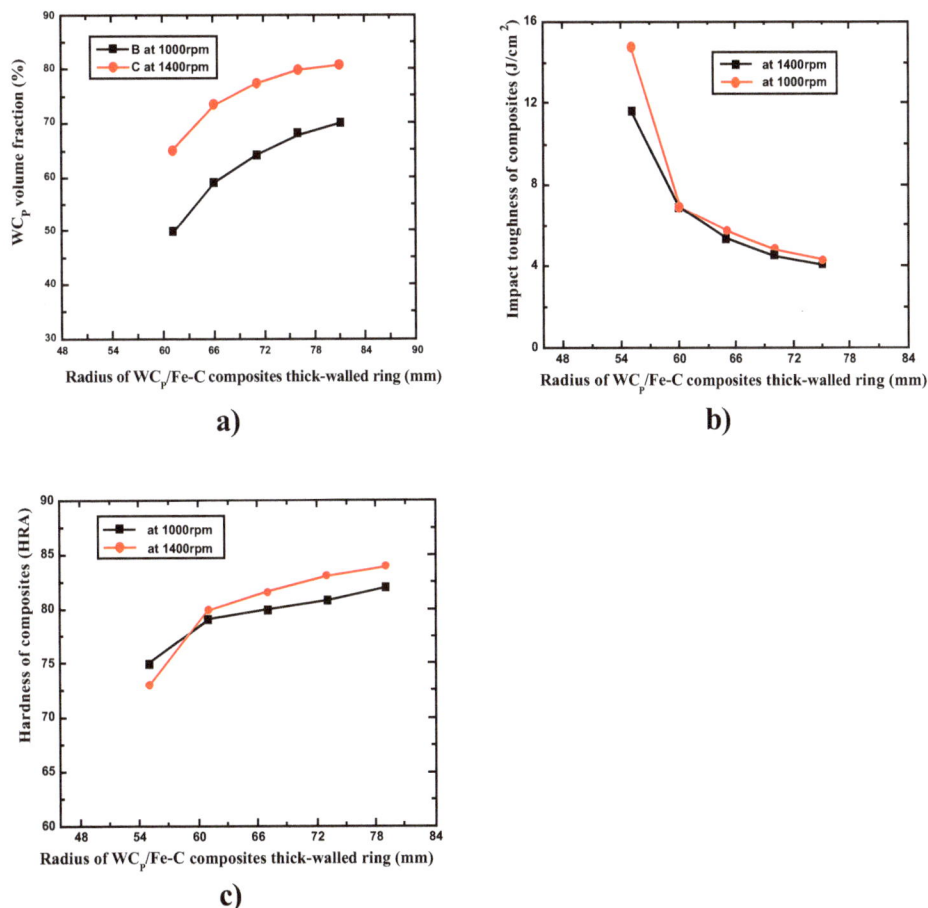

a)   b)

c)

Fig. 5.   Composition and mechanical properties along the radius at different rotational velocities: a) particle distribution, b) impact toughness and c) hardness (Ref. [19])

Another example of the obtaining of FGM is given by the centrifugal casting of aluminum alloy A356 reinforced by 5.7 vol.% nickel coated graphite particles of average diameter 75-150 μm. The graphite particles are introduced into the melt by a "know how" procedure and before this they are heated at temperature of 200°C for 2 hours. The initial mold temperature is 200°C, the melt temperature is 740°C, the rotation frequency is 1200 rpm and the casting size is 100 mm. In this case four regions are formed, as be seen in Fig. 6. The microstructure in the regions is shown in more detail given in Fig. 7.

Region I (a-d), which solidifies initially, does not contain graphite particles and can be characterized by a fine homogeneous dispersed phase Al_3Ni that is formed in the melt from the graphite particle coating. Region II (e-f) is also particle free region. There are some single particles there and eutectic is present in a substantial amount. Region III (g) is a transitional region between II and IV. Region IV (h-j) is a particle-rich region, which solidifies last. All the particles are packed here, due to centrifugal force.

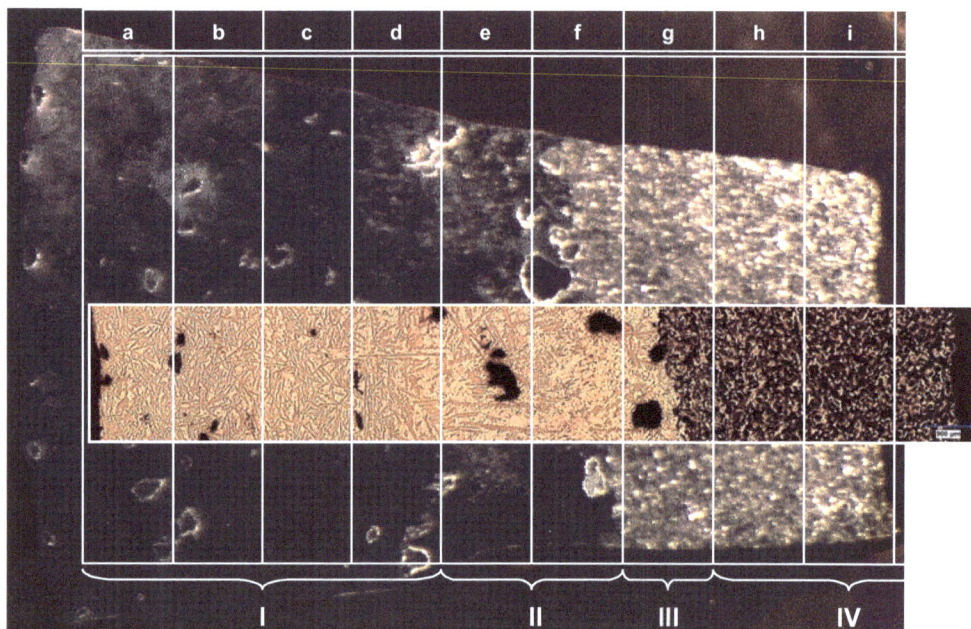

Fig. 6.   Macrostructure of MMC A356/5.7 vol. % Ni coated graphite particles obtained by centrifugal casting. In this cross section can be seen (from left to right): outer particle-free region of fine structure (I), middle particle-free region of more coarse structure (II), transitional area between particle-free and particle-rich region (III), and particle-rich region (IV)

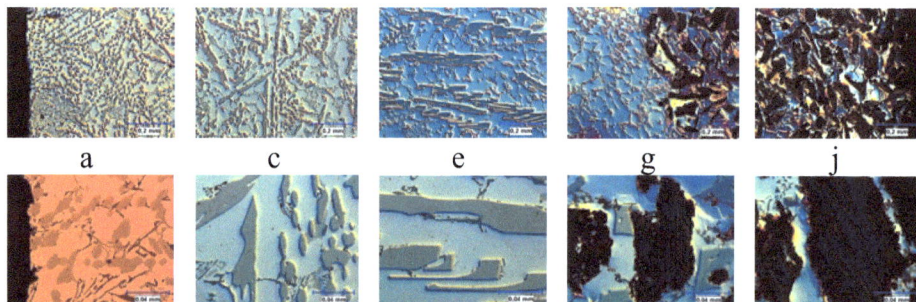

Fig. 7.   Microstructure of MMC A356/5.7 vol. % Ni coated graphite particles obtained by centrifugal casting. The sequence of images is denoted according to Fig. 6 and starts from outer casting region I (a, c), through the middle regions II (e) and III (g), to the inner region IV (j)

Centrifugal casting of MMCs, which consists of copper alloy C90300 (CuSn8Zn) and 7.5 vol.% graphite particles of average size 5 μm results in a casting that contains four regions in its cross section, Fig. 8. Markers on the cross section are referenced to Fig. 9. The graphite particles are introduced into the melt by a "know how" procedure. The initial mold temperature is 200°C. The melt temperature at which the particles are introduced into it is 1200°C, and the pouring temperature is 1120°C. The rotation frequency is 1000 rpm and the casting size is 100 mm. The microstructure in the regions is shown in more detail in Fig. 9.

In this case, the region I (a-h) contains a very small quantity of fine graphite particles and fine porosity. The region II (i-j) contains a relatively homogeneous distribution of graphite particles, crystals of metal matrix and some intermetallic components (TiC$_x$), which are related to "know how" used in the composite melt preparation. The particle rich region III (k-n) contains graphite particles and agglomerates.

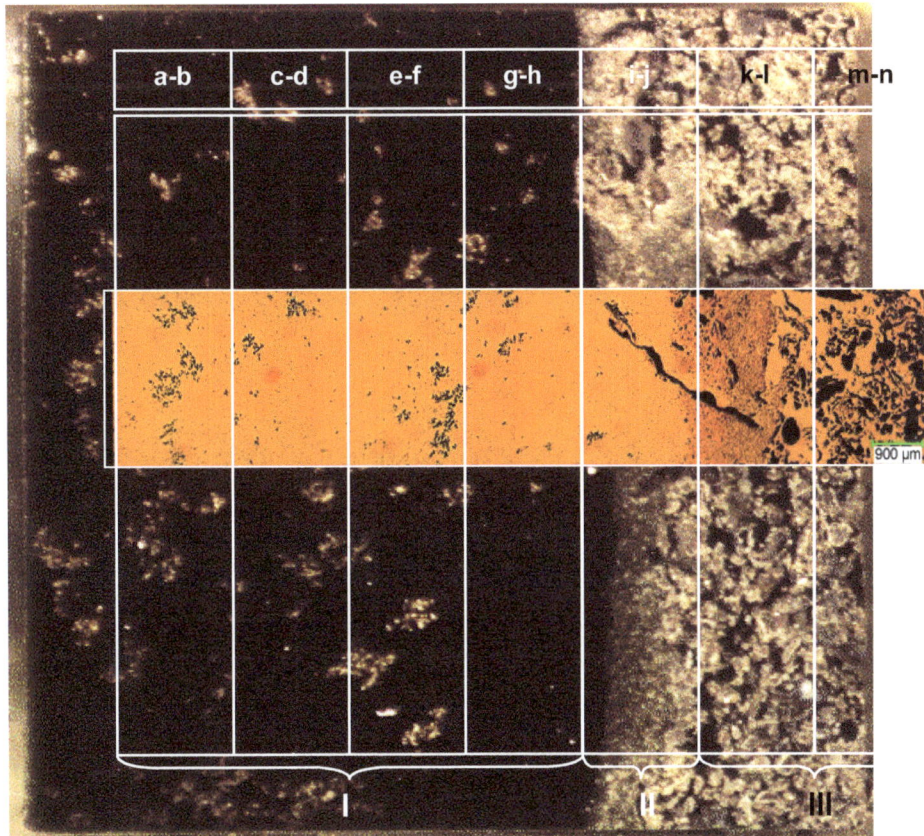

Fig. 8. Macrostructure of centrifugally cast MMC C90300/7.5 vol. % graphite particles. This crosssection contains (from left to right): outer particle-free region of fine structure and some quantity of micropores (I), transitional area between particle-free and particle-rich region (II), and particle-rich region (III)

Due to centrifugal force, solid particles that are of density lower than the melt move to the inner parts of casting part and solid particles of density higher that the melt move to its outer part. Because of this, hybrid MMC, which contain two or more types of reinforcing particles, can be obtained. Such a graded structure of specific reinforcing phase distribution based on aluminum matrix composite is given in Fig. 10. Two types of particles are used: SiC, 15 vol. % and 5 vol. % Ni-coated graphite. Two particle-rich regions of graded particle concentration are well-formed in the inner (graphite) and outer (SiC) parts of the casting.

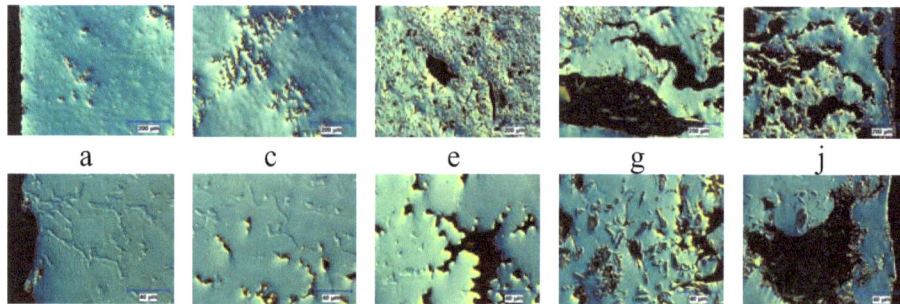

Fig. 9.   Microstructure of MMC C90300/7.5 vol. % graphite particles. The regions are denoted according to Fig. 8

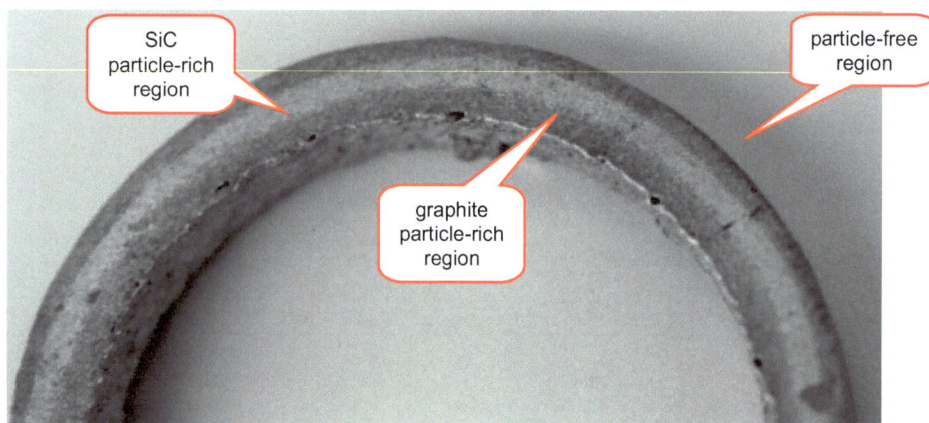

Fig. 10. Cross section perpendicular to the rotational axis of centrifugal cast MMC based on aluminum alloy A356 and 15 vol. % SiC particles and 5 vol. % Ni-coated graphite particles

I.4.   Structure Management by Magnetic or Electromagnetic Fields

Magnetic and electromagnetic fields are widely used in liquid metal processing. Many papers have been devoted to this matter, but here only the effect of a magnetic field on particles moving in liquid slurry need to obtain FGMs will be discussed. Generally, there are two lines of magnetic field applications: first, action on ferromagnetic particles in the production of reinforced ceramics by liquid routines, and second, generation of Lorentz force in a liquid MMC in order to affect particle movement during solidification of the liquid matrix.

A technology for production of ceramic-metal composites with a gradient of metal particle concentration is discussed in [20, 21]. Graded composites have been obtained by slip casting. The gradient of iron concentration is induced by magnetic field. Microstructures of the specimens are investigated qualitatively and quantitatively.

A technique for production of *in-situ* multilayer FGMs by the electromagnetic separation method using an alloy with two or more kinds of particles formed in different temperature ranges is presented in [22]. Threelayer FGMs are produced using an alloy of nominal composition Al-22 wt. % Si-3.9 wt. % Ti-0.78 wt. % B. These layers contain: first, ternary intermetallic compound Al-Si-Ti particles; second, primary Si particles; and third, Al-Si eutectic composition. During solidification the melt is subjected to Lorentz force, see Fig. 11.

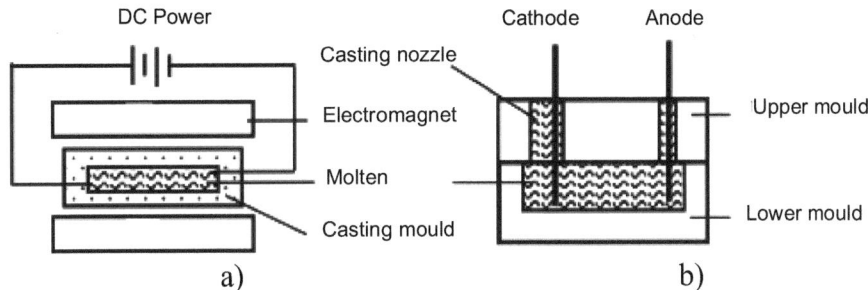

Fig. 11.   Diagram of equipment for the manufacture of FGMs by application of Lorentz force: a) top view and b) front view (Ref. [22])

The reason for the formation of the layers is that the intermetallic Al-Si-Ti particles, the primary Si particles and the Al-Si eutectic structure are formed in different temperature ranges and are affected by Lorentz forces of different magnitudes, Fig. 12. The interfaces between layers are gradual in respect to the microstructure, which is shown in Fig. 13.

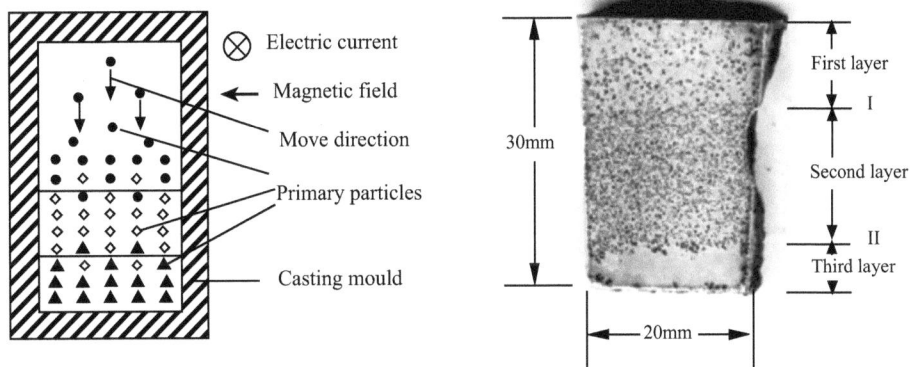

Fig. 12.   Particle movement in field of Lorentz force (Ref. [22])

Fig.13.   Ficrostructure of FGM obtained in-situ (Ref. [22])

The effect of processing parameters on particle distribution in Al/Mg$_2$Si FGMs obtained *in situ* by the same method is studied in [23]. The experimental results show that Lorentz force, solidification rate and alloy composition have great effect on the final particle distribution. Analysis suggests that these process parameters affect particle velocity and the total volume fraction of primary Mg$_2$Si particles. It is found that greater Lorentz force results in reduction of the particle-rich region and, conversely, smaller Lorentz force increases the width of the graded region.

Investigation of Al$_3$Ni phase distribution in *in-situ* obtained Al/Al$_3$Ni FGMs under Lorentz force can be found in [24]. Three primary alloy compositions are studied: Al-12 wt. % Ni, Al-17 wt. % Ni and Al-23 wt. % Ni. The concentration of the intermetallic component in the ingot volume is given in Fig. 14. In the region of higher particle volume fraction,

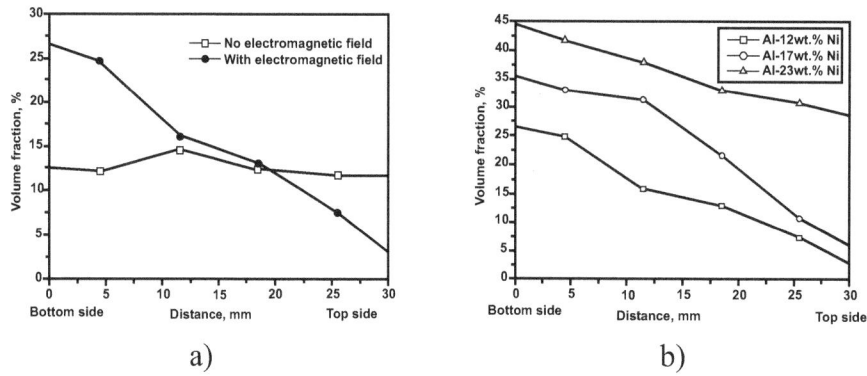

Fig. 14. Particle distribution in Al/Al₃Ni FGMs obtained under Lorentz force. a) with and without application of the force; b) distribution for different alloy composition (Ref. [24])

The major axis of Al₃Ni particles tends to be perpendicular to the Lorentz force, see Fig. 15. It is mentioned that the size of Al₃Ni particles becomes smaller in the case of solidification under Lorentz force compared with solidification without this force.

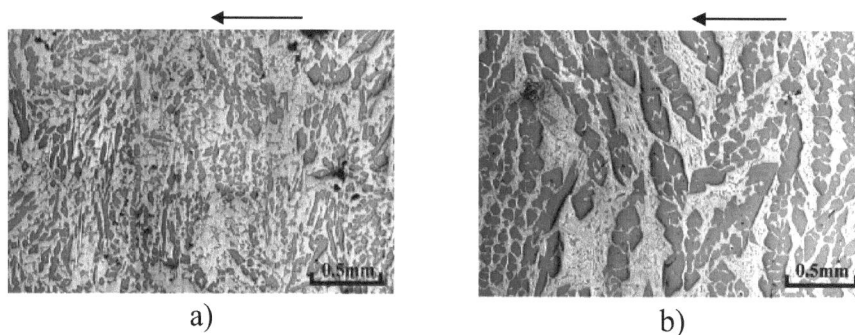

Fig. 15. Microstructure of particle rich region in the case of: a) Al-12 wt. % Ni primary alloy and b) Al-23 wt. % Ni primary alloy. The arrow indicates the direction of the Lorentz force (Ref. [24])

Application of Lorentz force during solidification of liquid metals and composite slurries is used not only for the generation of graded structure. In [25] there is presented a method for production of Al-Bi bearing alloy and method for its continuous casting in the presence of this force, which ignores gravity and facilitates formation of homogeneous structure. A separation system including a DC electric field and constant magnetic field is adopted for removal of inclusion in a molten Al alloy [26]. But very small inclusions, less than 10 μm, cannot be directly eliminated. To do this, small inclusions are first agglomerated by low-frequency electromagnetic vibration and are then removed by the same process.

A short discussion on the nature of the Lorentz force and its application to the manufacture of graded structures in MMCs is given in Appendix C.

I.5.   Functionally Graded Materials Obtained by Sintering

One of the methods often used for obtaining FGMs is *Spark Plasma Sintering* (SPS). The possibilities and limitations of this method are analyzed in [27].

It is reported for preparation of Cu/Al$_2$O$_3$/Cu [28], Ni/Al$_2$O$_3$/Ni [29], TiB$_2$/AlN/Cu [30, 31] FGMs and AlN-based composites [32] by the SPS method. Graded Cu/AlN/Cu composites are also manufactured [33] using SPS method and applying a two-step process. First, a symmetrical porous graded AlN plate is prepared using AlN powder consisting of particles of varying sizes. Afterwards, graded Cu/AlN/Cu samples are made by introducing Cu into the pores of the external porous AlN layer. The last materials are used for production of electrodes for thermoelectric generators.

Nano WC/Co hardmetals prepared by different SPS processes are manufactured and their properties are determined in [34]. A 4-layer FGM is also obtained by SPS, starting from powders of nano WC/10 wt. % Co, nano WC/12 wt. % Co, micro WC/15 wt. % Co and stainless steel disk. The other 3-layer FGM is made from powders of nano 21 wt. % Al$_2$O$_3$/ZrO$_2$, nickel and stainless steel. The SPS processing leads to FGM free of internal stress, which is measured using Vickers indentations.

Manufacture of Ti/Al$_2$O$_3$ composites by SPS is discussed in [35], by the same method, a Ti-Mg composite with graded density at low temperature can be made [36]. It seems that SPS can be considered a promising technique for obtaining graded structures in MMCs.

A specific design of Al$_2$TiO$_5$/Al$_2$O$_3$ system FGM is presented in [37]. Samples with various volume contents of Al$_2$TiO$_5$ are prepared by powder metallurgy sintering. The basic material properties of the samples are then tested and used as a basis for predicting the material properties of the transient layers in Al$_2$TiO$_5$/Al$_2$O$_3$ FGMs. The thermal stresses produced in the fabrication process are simulated by using a finite element method. On the basis of the comprehensive considerations of the minimum residual thermal stress and distribution state, the optimum compositional distribution in Al$_2$TiO$_5$/Al$_2$O$_3$ FGMs is achieved.

A production method for metal components with microsized porous structure is developed by applying "powder space holder method" to a metal powder injection molding process [38], Fig. 16.

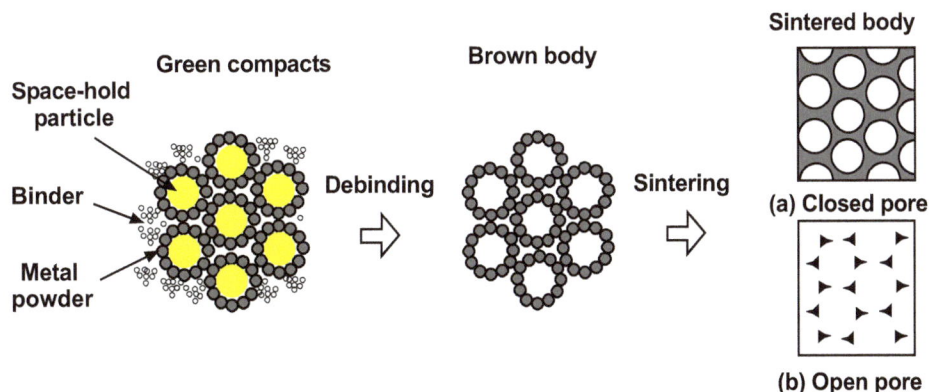

Fig. 16. Conception of micro-porous metal injection molding realized by the powder space holder method, (Ref. [38])

The co-sintering process is utilized to make a plate of sintered metal with micro porous graded structures, Fig. 17.

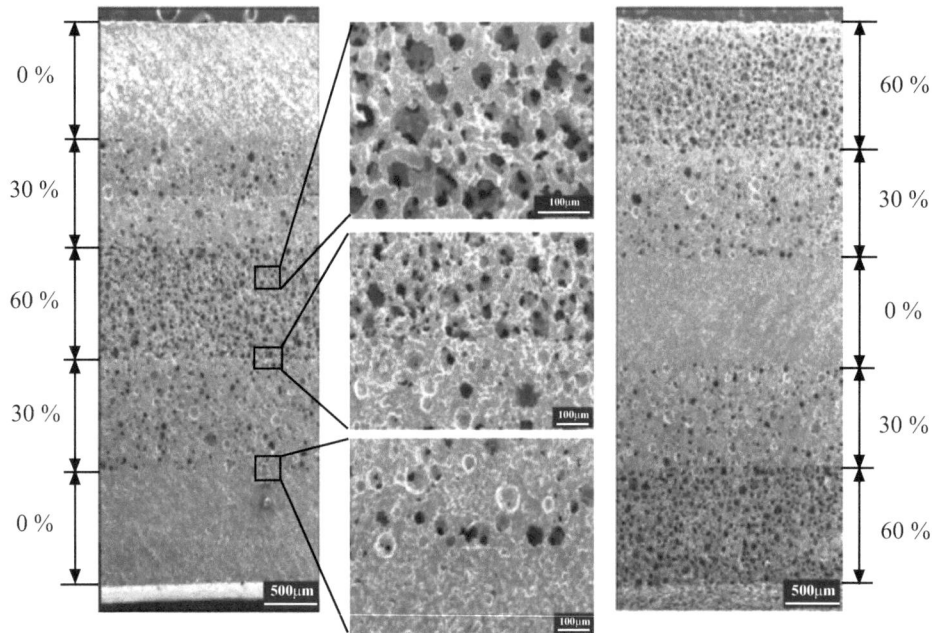

Fig. 17. Structure of porous graded metals produced by co-sintering of laminated stacked green compacts with various percentages of space-hold particles (Ref. [38])

The green compact sheets with various contents of space holder particles are prepared by hot press molding for simplicity. The five layers of metal with symmetric structure, in which the skin layer is formed with high density metal and the core is formed with open or closed porous structure, or with inverse symmetry, are obtained by changing the stacking sequence in the co-sintering process. Mechanical properties of the materials with plain homogeneous porous structure and porous graded structure are compared. The usefulness of proposed method for production of metal components with micro porous graded metal structures is shown.

A production technique which consists of tape casting and sintering, for a novel class of electroconductive ceramics in which the conductive phase is molybdenum disilicide composites with AlN-SiC-$MoSi_2$ at 55/15/30 and 60/20/20 vol. % concentration respectively, is presented in [39].

In order to reduce the production cost of WC alloy and give wear resistance to austenitic stainless steel, Kawakami et al. [40] apply *Pulsed Current Sintering* (PCS) method and produce some specific FGMs. Mechanical properties of these materials are evaluated by a threepoints bending test, and microscopy observations of their microstructures are made. Sintering conditions are optimized in order to obtain functionally graded layers.

Functionally graded materials produced by sintering of commercially available partially stabilized zirconia (ZrO_2 - 3 mol. % Y_2O_3, PSZ) and austenitic stainless steel (SUS 304) powders are studied in [41]. The fracture toughness is determined by conventional tests for several non-graded composites (non-FGMs) and by a method utilizing stable crack growth for FGMs. Based on the experimental results, the fracture mechanism and influences of microstructure on fracture toughness distribution, as well as difference in fracture toughness between the

FGMs and non-FGMs, are discussed. The influence of microstructure on the fracture toughness of the non-FGMs is negligible, while the fracture toughness on the FGMs is higher in fine microstructure than in coarse microstructure. The fracture toughness in the FGMs is higher than that of the non-FGMs especially in the case of fine microstructure.

Cho and co-workers [42] manufacture graded composites consisting of many (up to nine) layers in an Al-SiC$_p$ system. At the beginning they prepare by sintering a multi-layered perform, Fig. 18.

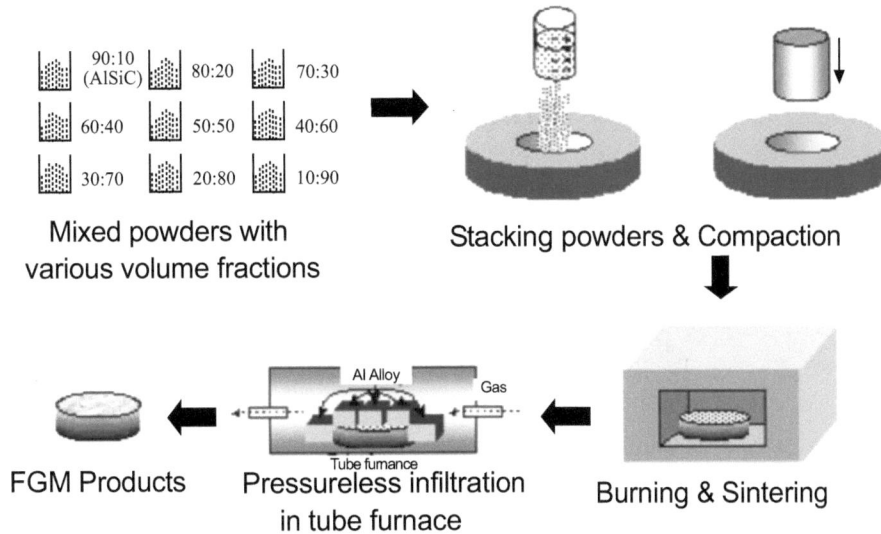

Fig. 18. Manufacturing process of Al-SiC$_p$ FGM (Ref. [42])

Each layer contains Al and SiC particles in a different ratio. The preform is heated in a furnace and infiltrated in pressureless conditions by molten aluminum alloy. The structure obtained is shown in Fig. 19.

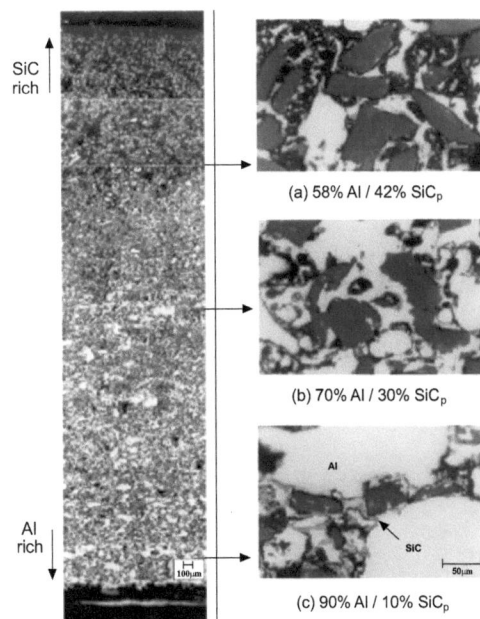

Fig. 19. Microstructure of Al-SiC$_p$ FGM (Ref. [42])

A theoretical model is used for analysis of the thermomechanical properties of these FGMs in [43]. It is demonstrated that the model can be applied to predict the internal stress distribution in Al-SiC$_p$ FGMs plates (3 mm), and thus can be used effectively to design multilayered FGM structures without any critical failure during their usage.

Dense FGMs with a gradually changing fraction of very hard and light B$_4$C particles are successfully fabricated and studied in [44] by an original reactive forging combustion synthesis method. A self-sustained exothermic reaction is ignited in a Ti-Ni-carbon powder blend at about 1000°C, which is slightly above the Ti$_2$Ni-Ti eutectic temperature, yielding a Ti$_2$Ni-TiNi-TiC matrix. The addition of 10 to 50 wt. % B$_4$C particles did not significantly change the combustion temperature that remained above 1600°C, i.e. well above the melting of Ti$_x$Ni$_y$ intermetallics. The presence of a liquid phase during combustion allows near full densification of the synthesized Ti$_x$Ni$_y$-TiC-B$_4$C composites with homogeneous and gradient distribution of B$_4$C particles under a moderate pressure of 200 MPa applied for 1 minute. A reaction zone several tens of microns thick is observed at the B$_4$C-matrix interface containing a large amount of fine TiB$_2$ particles. Precoating B$_4$C particles by PIRAC annealing in Ti powder prior to processing allows significant reduction of the matrix reinforcement interaction during combustion. Two of the structures obtained are shown in Fig. 20. The above results suggest that reactive forging of exothermic powder blends with gradually changing fractions of B$_4$C or other hard particles is a feasible path for manufacturing dense functionally graded materials.

A new resistance sintering method under ultra high pressure for manufacture of Mo/Cu FGM is presented in [45]. The consolidation is carried out under pressure of 8 GPa and with on input power of 15 kW for 50 s. The structure obtained is shown in Fig. 21. Micro-hardness and bend strength are measured to evaluate the sintering effects. The densification effect and microstructure of these FGM samples are investigated and the sintering mechanism is discussed. It is found that the micro-hardness and bend strength of the middle layers of FGM increase as the Mo content increases. It is concluded that metals with a large difference in melting points (such as W/Cu, Mo/Cu composites) can be well sintered by resistance sintering under ultra-high pressure.

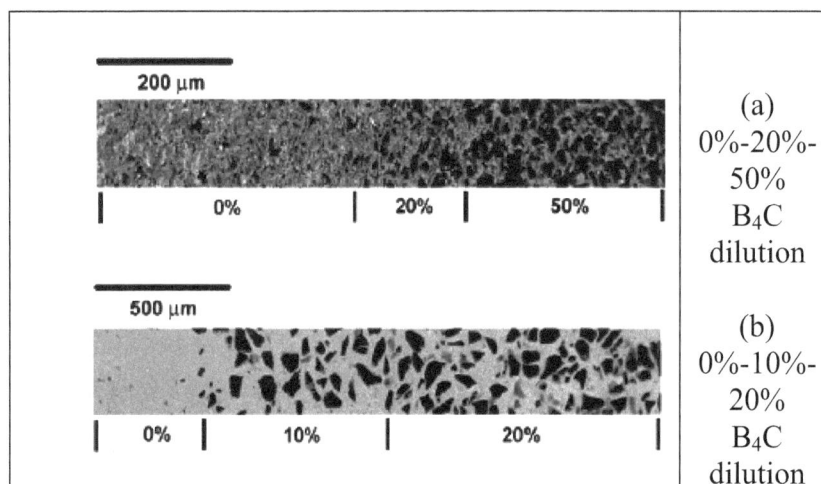

Fig. 20.   SEM micrographs of triple-layered FGMs with: a) 0% - 20% - 50% B$_4$C and b) 0% - 10% - 20% B$_4$C in 1.4Ti - 0.5C - 0.4Ni matrix (Ref. [44])

Fig. 21. SEM micrograph of well-sintered Mo/Cu FGM (Ref. [45])

Brinkman at al. [46] developed based on *powder metallurgy,* a reactive hot-pressing method in which the material can flow during the transient liquid stage. This method is used for obtaining Al-TiB$_2$ composites, which are characterized by high volume fraction of TiB$_2$ close to the ingot surface and decreasing concentration of TiB$_2$ inside the ingot volume. The synthesis is based on the reaction

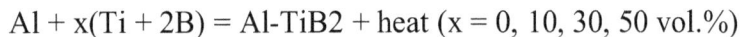

$$Al + x(Ti + 2B) = Al-TiB2 + heat \ (x = 0, 10, 30, 50 \ vol.\%)$$

The authors conclude that FGMs can be produced by this method but control over the resulting structure is very difficult.

Production of Mo-Si$_3$N$_4$ FGMs with 6 layers by hot-pressing of powder mixtures and their application as thermal barrier materials is discussed in [47].

A short overview on powder metallurgy of tungsten alloys is given in [48]. Techniques for the production of W-Cu FGMs are also discussed.

In [49] the authors report the results of investigations on sintered FGMs of Al$_2$O$_3$ and ZrO$_2$ and their densification as a function of the starting suspensions' properties (alumina/zirconia ratio, solid content and dispersion state), with the aim of compensating for different amount of sintering shrinkage.

I.6.   Infiltration into Graded or Homogeneous Preforms

A method for production of open-pore microcellular aluminum by replication processing is presented in [50]. The chart in Fig. 22 represents the schematically method. The method also makes possible the production of functionally graded material structures, by selective placing of salt preforms having various densities or, alternatively, continuously varying density.

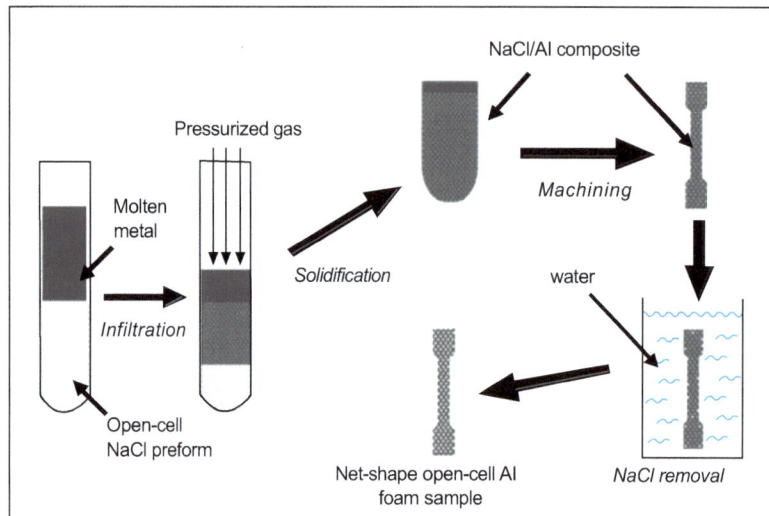

Fig. 22. Schematic description of the replication process as used for the production of open pore microcellular aluminum (Ref. [50])

Such examples can be seen in Fig. 23, including a five-layer sandwich beam featuring two dense aluminum outer skins, and an open-pore aluminum foam core consisting of three discrete layers: a central portion of relatively high density and layers of lower density adjacent to the dense skins.

A technique that allows the manufacture of continuous fiber reinforced MMC with graded fiber content is presented in [51]. Carbon fiber is used as the reinforce phase and magnesium alloy AM20 (2 wt. % Al, 0.6 wt. % Mn, balance Mg) is used as the matrix. Infiltration in the graded preform is realized under gas pressure. Mechanical tests are carried out on these MMCs. It is concluded that the bending strength, the tensile strength, the stiffness and the thermal expansion can be tailored by changing the fiber/matrix ratio.

Production of tungsten/copper FGMs by copper infiltration in a sintered tungsten graded preform is performed in [52]. The profile of the gradient is predicted with a mathematical model. Young's modulus and the thermal expansion coefficient of the materials are measured.

A method for production of Cu/W FGMs is discussed in [53]. The first step of this method involves preparing a graded tungsten preform by vibration of W agglomerates in order to obtain skeleton with a gradient in porosity. After this the preform is sintered without pressure and infiltrated with molten Cu. The final Cu-infiltrated FGMs are characterized microstructurally and their electrical resistivity is analyzed.

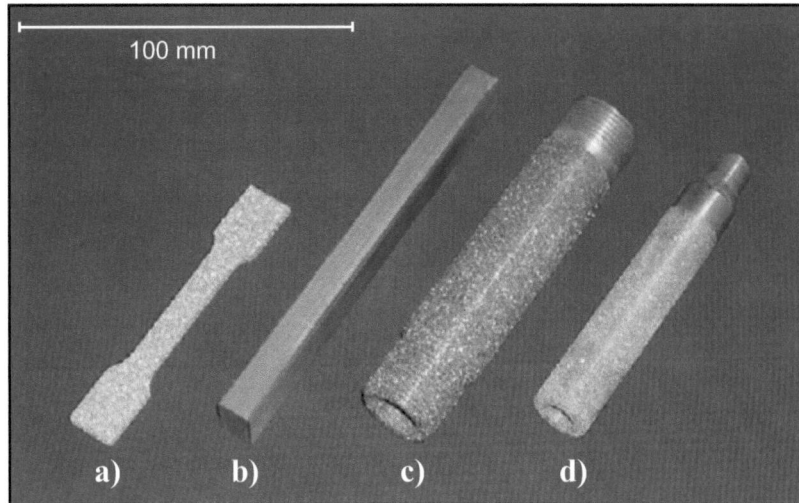

Fig. 23. Cast and machined metal foam specimens: a) flat dog-bone tensile test specimen, b) five-layer sandwich beam featuring two dense aluminum outer skins, two intermediate layers of lower-porosity metal foam and a central higher porosity metal foam core, c) machined filter structure with two tapped end threads (one in dense Al, one in the metal foam) and d) net-shape cast filter structure with an as-cast male thread in a dense metal end and an as-cast female thread in the other (metal foam) end. a), c) and d): 400 μm average pore size, b): average pore size 75 μm (Ref. [50])

An apparatus for computer controlled pressure filtration has been constructed and described in [54]. The precise flow rate of multiple ceramic suspensions of different compositions can be controlled. Suspensions are mixed before entering the filtration chamber. By dynamically adjusting the different flow rates, phase compositions between 100 % phase A and 100 % phase B can be realized as a function of the filter cake height with filtration pressures of up to 10 MPa. This is suitable for producing functionally graded materials with smooth changes of composition or layered structures where steep-like changes are necessary. Gradients and layers in porous alumina (Fig. 24) and in alumina/zirconia (Fig. 25) have been produced and investigated. The process is suitable for all materials that can be handled by slip casting.

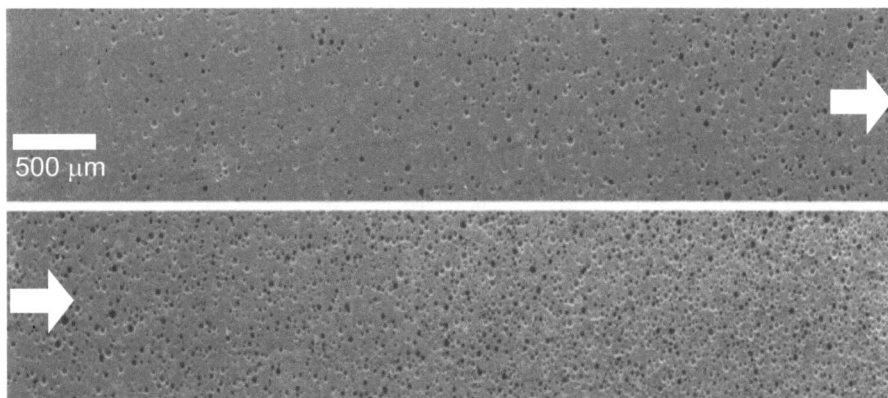

Fig. 24. Alumina with a continuous porosity gradient showing variation between 0 and 50 % porosity obtained by burning out of pore forming agents (Ref. [54])

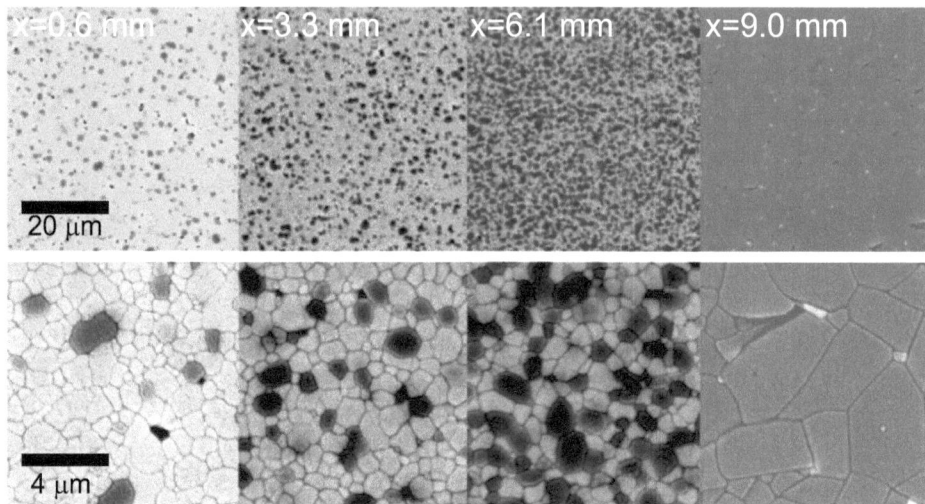

Fig. 25. FGM sample microstructures of linear alumina/zirconia gradient taken across the sample thickness as indicated by the coordinate x (Ref. [54])

I.7.   Solidification of Graded Slurries

Sahm and Guentner [55] used a *multi-pouring method* with controlled mold filling for production of FGMs of different types of aluminum alloys and with SiC particles as reinforce phase. They realized controlled mold filling of two melts in a gravity casting and centrifugal casting, see Fig. 26. The width of the graded interface in the castings strongly depends on the solid fraction of the first cast alloy formed before pouring of the second alloy. This also depends on the casting geometry.

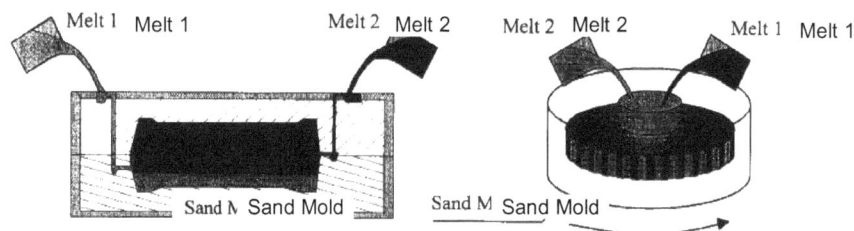

Fig. 26. Multi-alloy casting processes: a) gravity and b) centrifugal (Ref. [55])

I.8.   Manufacture of Graded Structured Coatings and Plates

Particle reinforced metal matrix composite coatings have been developed in view of property combinations such as increased hardness, high creep/fatigue resistance as well as superior wear and oxidation resistance. MMC coatings can be produced by various techniques, which include powder metallurgy, liquid metal processes and *Electrophoretic Deposition* (EPD).

Gradient layer structured coatings are extremely useful in the environmental protection of metal structures at high thermal and mechanical loadings. A comprehensive overview of properties, production technologies and future trends in this field is presented in [56].

Functionally graded structural ceramic/metal composite coatings for relaxation of thermal stress, functionally graded anti-oxidation coatings for carbon/carbon composites, and functionally graded dielectric ceramic composites to develop advanced dielectric ceramics with flat characteristics of dielectric constant in a wide temperature range are elaborated in [57]. This paper introduces functionally graded coatings for C/C composites with superior oxidation resistance at high temperatures.

Preparation of graded Al/AlTi and Al/Al-Si coatings by deposition of nanoparticles with a *Supersonic Free-Jet* (SFJ-PVD) is described in [58]. The method is developed as a new coating method in which a coating film is formed by depositing nanoparticles with very high velocity onto a substrate. The high velocity of nanoparticles is produced by the supersonic gas flow of inert gas. A smooth, compact and defect-free microstructure is formed both at the interface between substrates and coating films and inside the coating films. The microstructures of Al/AlTi and Al/Al-Si coating films have a very fine grain size. Mixing Ti and Al nanoparticles by depositing them onto a substrate produces *in-situ* syntheses of γ-TiAl and γ_2-Ti$_3$Al intermetallic compounds on the substrate. It is confirmed with nano-indentation hardness tester that the graded coatings have graded hardness corresponding to the gradation of their composition.

Deposition of diamond-like carbon based functionally graded Ti/Ti$_x$C$_y$/DLC coatings by hybrid technique of magnetron sputtering and DC plasma-enhanced chemical vapor deposition is performed in [59]. The coatings parameters and their mechanical properties are investigated.

Electrophoretic deposition is a technique more often applied to ceramic materials. In this technique a stable suspension of powders is used in which the powder particles have acquired an electric charge through interaction with the suspension medium. The electric charge causes migration of the particles under the influence of an electric field introduced in the suspension (electrophoresis). Under certain circumstances the particles will move to the electrode with opposite charge, lose their surface charge and form a deposit on this electrode (electrodeposition), see Fig. 27a). The deposited layer can be several millimeters thick, and by this technique not only coatings but also plates can be produced.

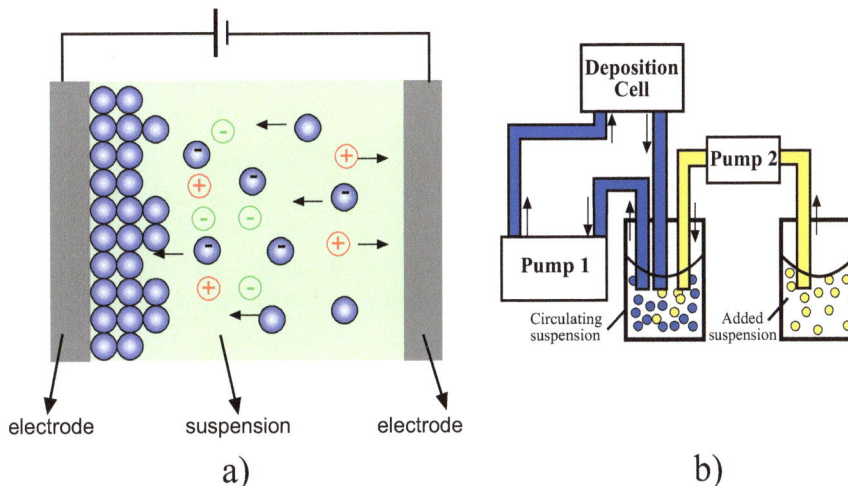

electrode   suspension   electrode

a)   b)

Fig. 27. Schematic representation of: a) the electrophoretic deposition process, and b) electrophoretic deposition system for production of FGM (Ref. [60])

Electrophoretic deposition allows the formation of FGMs by deposition from a powder suspension, to which a second suspension is continuously added during the process, see Fig. 27b). A discussion on the main features of this method is given in [60]. A technology for production of graded WC-Co plates is also presented in that work.

Production of homogeneous and graded ZrO_2-WC composites with up to 55 wt. % WC by means of EPD is reported in [61]. The EPD processing parameters and the composites' mechanical properties are measured and evaluated. A mathematical model of the EPD process is developed and used for calculating the composition gradient in the ZrO_2-WC FGM, which is a function of starting composition of the suspension, the EPD operating parameters and the powder specific EPD characteristics.

Further details about the EPD method, structures and properties of the graded composites obtained can be found in [62].

Combined techniques for the manufacture of tailored powder composites with specially distributed pore-grain structure and chemical composition are presented in [63]. Electrophoretic deposition followed by microwave sintering is employed to obtain FGMs by controlling *in-situ* the deposition bath suspension composition. Al_2O_3/ZrO_2 and zeolite FGMs are successfully synthesized using this technique. In order to produce an aligned porous structure, unidirectional freezing followed by freeze drying and sintering is employed. By controlling the temperature gradient during the freezing of powder slurry, a unidirectional ice-ceramic structure is obtained. The frozen specimen is then subjected to freeze drying to sublimate the ice. The obtained capillary-porous ceramic specimen is consolidated by sintering.

A study of EPD applied for obtaining Ni-Al_2O_3 composites is presented in [64]. The process conditions are optimized and gradient Ni-Al_2O_3 composites are prepared. Microstructural characterization and hardness measurements are performed. The effect of process parameters on the amount and distribution of Al_2O_3 particles as well as the morphology of the deposits with and without Al_2O_3 gradients are presented and discussed.

A precise electrophoretic deposition as a near net shaping technique for production of functionally graded biomaterials is described in [65]. Production techniques, structures and mechanical properties of metallic and ceramic FGMs for medical application as biocompatible materials are widely discussed in [66-71].

Among deposition methods for graded coating production, *Pulsed Laser Deposition* (PLD) is considered as a unique tool for preparing high quality surface mono- or multilayers. Deposition procedures, design, microstructure and tribological properties of Cr/CrN and Ti/TiN based coatings are widely discussed in [72].

I.9.   Graded Structures in Ordered Porosity Metal Materials (Gasars)

A special class of metal porous materials, which appears as gas reinforced metal matrix composites, consists of gasar (lotus type or ordered porosity) materials. They are obtained by unidirectional solidification of a gas super saturated melt. Due to higher gas solubility in the liquid phase, solidification of the metal and nucleation of gas pores occur simultaneously, which results in the formation of an ordered gas-solid composition. This phase transformation is very similar to the conventional eutectic reaction, but one of the phases in the resulting structure is a gaseous phase. The major advantages of gasars over other porous materials are:

- improved strength and rigidity;
- the possibility of making regular or graded structures;
- wide range of pore diameter (10 μm to 10 mm);
- feasible control of pore shape and orientation;
- easy manufacture and relatively low cost.

Like all porous metals, the ordered porosity metals have a wide range of applications including filters, catalysis, silencers, flame arresters, heat exchangers, fuel cells, electrolytic cells, fluid substance separators, ionic rocket engine parts, self-lubricating bearings, thermal screens and vibration dampers. The most attractive applications are as components of rocket combustion chambers and oil high pressure filters [73]. Gasar stainless steel is a promising material for medical applications, especially for artificial bones or joints with good corrosion resistance [74].

Originally these materials were cast in an open metal mold located in a chamber of high gas pressure, Fig. 28 a). Such a method is suitable for Cu, Al and its alloys with high thermal conductivity. Gasars fabricated by this technique from metals and alloys of relative low thermal conductivity do not have uniform pore size and porosity [75] because the solidification rate becomes lower as the solidification proceeds, which results in the pore size increasing. In order to ensure flexible process control, new techniques are being developed [76]. During the last few years Japanese scientists have applied specific modification of the continuous zone melting technique [77], Fig. 28 b) and continuous casting method [78], Fig. 28 c). These techniques provide flexible control of the structure formation process and produce a great variety of metallic and nonmetallic porous materials.

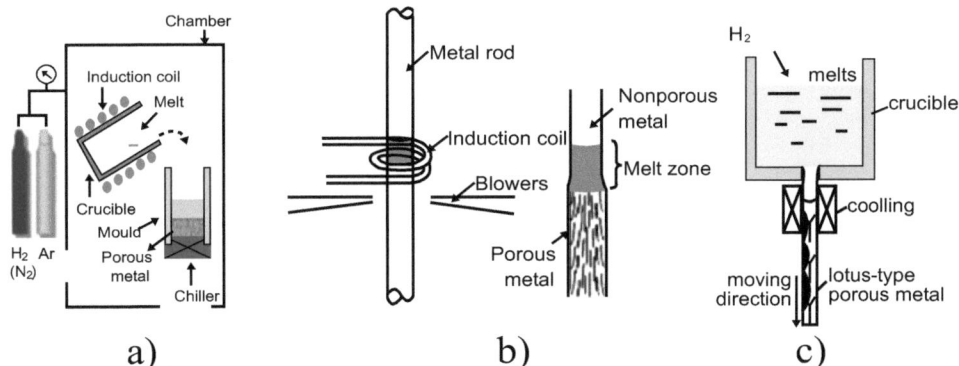

Fig. 28. Schematic presentation of technologies for gasar ingot production: a) open mold casting (Ref. [79]), b) continuous melting techniques (Ref. [77]) and c) continuous casting method

(Ref. [78])

Gasar structure is controllable by verifying the processing parameters (gas pressure during melt saturation and solidification, initial melt and mold temperature, cooling rate, direction of heat removal, etc.) and a large variety of structures can be obtained, see Figs. 29-34. To do this in a more effective way it is highly recommended to use mathematical models of separate phenomena or a complex mathematical description of entire technological process. An attempt at a comprehensive mathematical description of structure formation in the gasar process and its potential for graded porous structure manufacture appears in Appendix D. Other simpler models have been published in [80, 81].

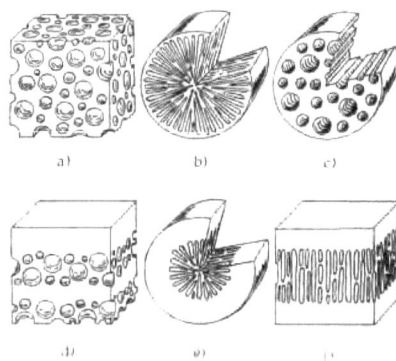

Fig. 29. Diagrams of pore morphologies potentially available with the gasar process: a) spherical,
b) radial,
c) cylindrical and
d-f) laminates (Ref. [82])

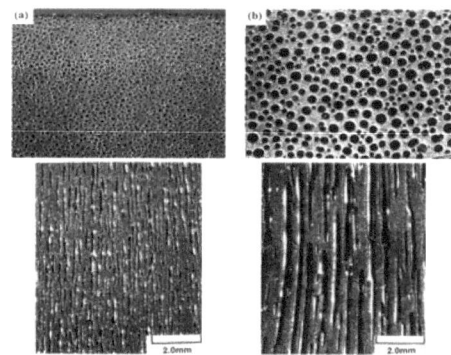

Fig. 30. Typical micrographic pictures of transverse (top) and longitudinal (bottom) cross-sections of copper gasar:
a) porosity 32.6 vol. %,
b) porosity 44.7 vol. % (Ref. [83])

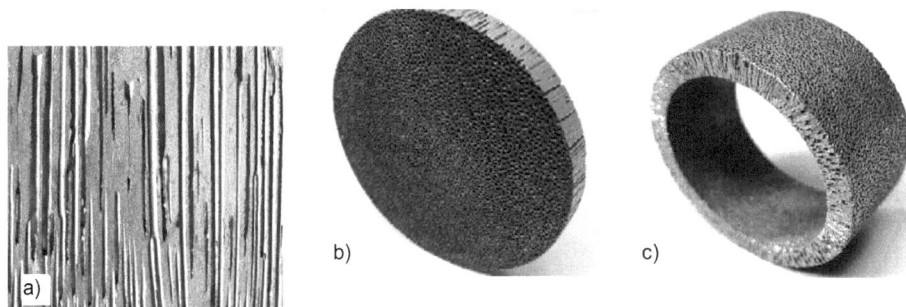

Fig. 31. Typical gasar structures: (a) parallel cylindrical pores in longitudinal section; (b) parallel ellipsoidal pores; (c) radial ellipsoidal pores (Ref. [84])

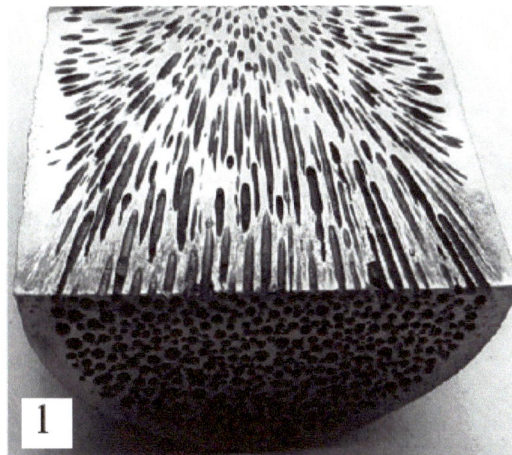

Fig. 32. Typical gasar structure fabricated by open mold casting. Combination of parallel and radial ellipsoidal pores (Ref. [73])

Fig. 33. Ordered porosity stainless steel fabricated by a continuous zone melting technique (Ref. [85])

Partial gas pressures applied on the melt before and during solidification are basic driving parameters for structure control. During ingot solidification the porosity can be managed in a wide range, from 0 up to 75 vol. % for some metal-gas systems. This allows the production of FGMs with specific characteristics. Changes to the partial gas pressures cause essential changes in pore nucleation and pore number per unit area. When the pressures are constant, ingot porosity can slowly increase in open mold casting through increase of pore diameter but not through a rise in pore number. A more flexible method for production of controlled gradient porous structures is continuous casting, Fig. 28 c).

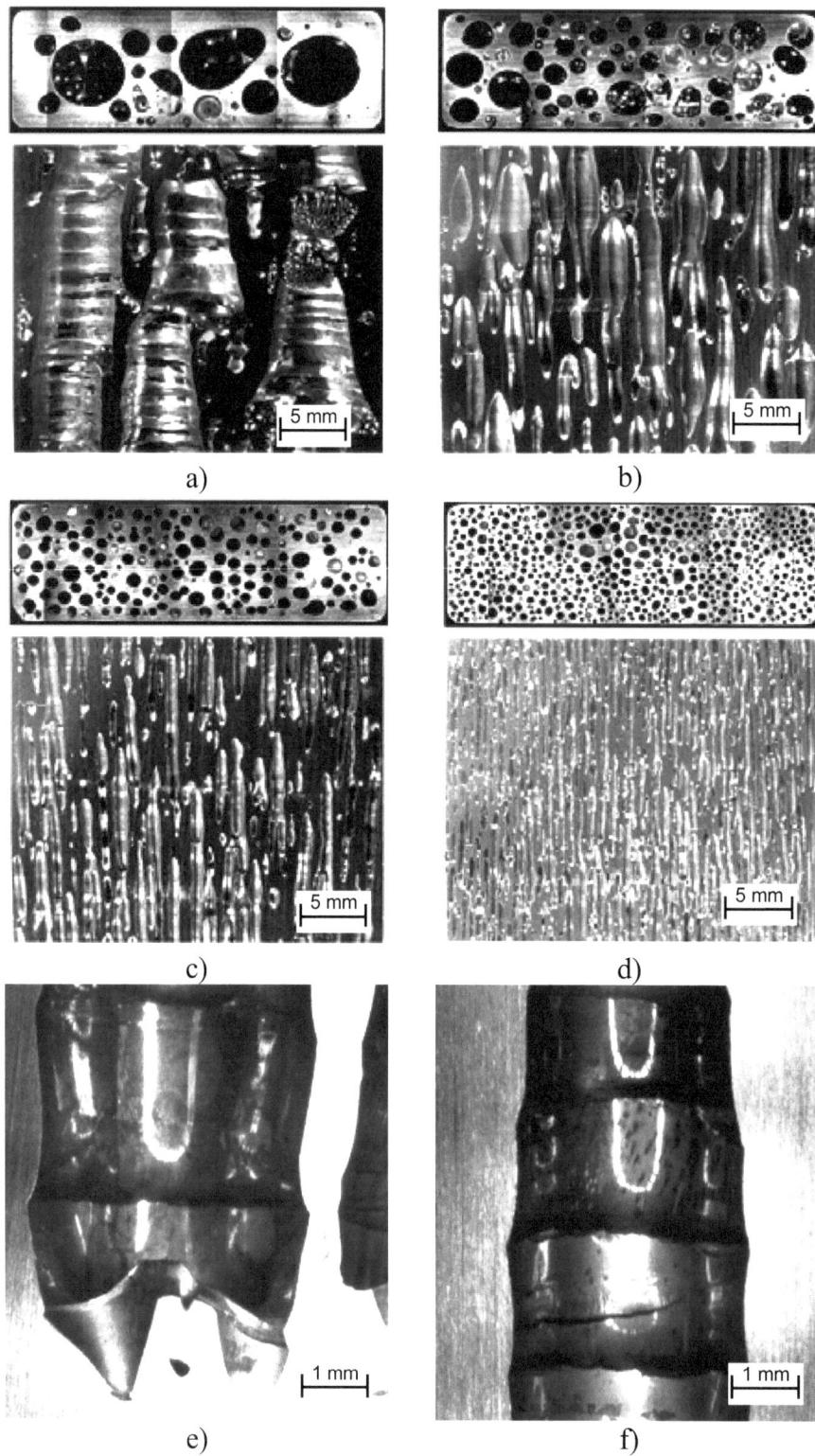

Fig. 34.   Cross sections of copper gasar made by continuous casting under a hydrogen gas pressure of 1.0 MPa. (a)-(d) perpendicular and parallel to transference direction, solidification velocity: (a) 1 mm/min, (b) 5 mm/min, (c) 10 mm/min, (d) 20 mm/min (e) starting part of a pore and (f) middle part of a pore (1 mm/min) (Ref. [86])

CHAPTER II: SPECIFIC PROPERTIES CHARACTERIZATION AND MATHEMATICAL MODELING

Ludmil B. Drenchev[1], Jerzy Sobczak[2]

[1]*Institute of Metal Science, 67, Shipchenski Prohod Street, 1574 Sofia, Bulgaria,* [2]*Foundry Research Institute, 73, Zakopianska St., 30-418 Krakow,Poland*

Address correspondence to: Ludmil B. Drenchev; [1]*Institute of Metal Science, 67, Shipchenski Prohod Street, 1574 Sofia, Bulgaria; Telephone: +359 2 46 26 223, Fax: +359 2 46 26 300, Email: ljudmil.d@ims.bas.bg*

Abstract: This chapter is focused on some specific characteristics of functionally graded materials that need special evaluation.

II.1. Properties Measurement

A great variety of particles (metallic, intermetallic, ceramic) are used as the reinforcing phase or initial materials in the manufacture of FGMs. It is extremely important to have a certain procedure for quantitative measurement of the complex of geometrical characteristics of these particles. Such a procedure, developed based on computer-aided methods, is proposed in [87].

Tribological tests of aluminum based matrix composites with functionally graded properties are presented in [88]. SiC particulate reinforced F3S.20S aluminum matrix composite (*Duralcan*) is melted and centrifugally cast in order to obtain a gradient particle distribution. The effect of particle size on graded particle distribution is analyzed in [89]. Friction and wear behavior of such composites (accompanied) with concurrent corrosion processes is studied in [5, 8, 90].

Thermal conductivity at various temperatures of Al-SiC_p graded composites is studied in [42]. It is found that conductivity of Al-SiC_p composites increases non-linearly with decreasing SiC volume fraction. Cyclic thermal shock fatigue tests were also performed.

X-ray microtomography research for evaluation of Al/SiC wetting characteristics in centrifugally cast FGMs is presented in [7]. This study facilitates understanding of microporosity formation and deterioration of mechanical properties in such composites.

In order to develop a non-destructive method for investigation of inhomogeneous microstructure of FGMs, J. Bonarski [91] successfully applies X-ray Texture Tomography (XTT). The application of XTT is demonstrated on examples of galvanized car body sheet steel and Si/HfN electronic composition.

The joining of FGMs based on the Al-2124/SiC_p composite system to each other using a rotary friction welding technique is described in [92].

II.2. Theoretical Approaches for Estimation of FGMs Properties

Metal matrix composites of the type Al/SiC or Al/Al₂O₃ contain significant residual stresses due to different thermal expansion coefficients type from those of the metal and ceramic constituents. They are believed to influence the mechanical properties of these materials to some extent, including changes in their failure behavior. A physically based micromechanical approach is developed in [93], in order to study the influence of residual stresses on local as well as global properties of MMCs.

Numerical crack analysis of two-dimensional FGMs is performed in [94], using a boundary integral equation method. Numerical examples are presented to explore the effects of the material gradients and crack orientation on the stress intensity factors.

A specific method for approximate solution of the heat transfer problem in FGM with arbitrary geometry of reinforcing particles is developed in [95]. Using this approach, the effective thermal conductivities for composites with different inclusion geometry are calculated. It is shown that the effective thermal conductivity depends not only on the volume fractions and the properties of components but also on the inclusion's geometrical parameters.

In [96] a three–dimensional model for surface cracking of graded coatings bonded to a homogeneous substrate is elaborated. The main objective is to model the subcritical crack growth process in the coated medium under cyclic mechanical or thermal loading.

The creep property of functionally graded materials (ZrO₂/Ni system) in a high temperature environment is investigated theoretically in [97] by a computational micromechanical method.

An efficient approach for theoretical investigation of the dynamic response of cracks in FGMs under impact load is presented in [98]. Analytical solutions of stress fields in functionally graded hollow cylinder with finite length subjected to axissymmetric pressure loadings on its inner and outer surfaces are presented in [99]. Numerical results for stresses in the cylinder at various of loads are presented.

The thermal fracture and its time-dependence in functionally graded yttria stabilized zirconia - NiCoCrAlY bond thermal barrier coatings are studied in [100]. The response of three coating architectures with similar thermal resistance to laser thermal shock tests is investigated experimentally and theoretically.

A micromechanics-based elastic model is developed in [101] for two-phase functionally graded composites with local pair-wise particle interactions. Averaged elastic fields are obtained for transverse shear loading and uniaxial loading in the gradient direction.

The process of multi-particle setting in the manufacture of FGMs by co-sedimentation is modeled in [102]. The models can be used to design powder composition and to predict the volume fractions obtained in FGMs. As examples,

TiC-Ni system FGMs are designed and manufactured. The predictions fit well the actual results obtained. An experiment with Mo - Ti system FGM is also used to validate the prediction model.

APPENDICES

Appendix A

GRADED STRUCTURES BY SEDIMENTATION AND FLOTATION

Sedimentation and flotation seem to be simple physical processes and it should be easy to make use of them. To do this in a real situation one must be very familiar with the processes. Employment of mathematical modeling and numerical simulations is extremely helpful here. A great number of pages have been devoted to this matter [102-108]. Some basic features of these phenomena, which are closely related to the manufacture of graded structures in particle reinforced MMCs, will be given below.

Sedimentation and flotation in the process of MMCs preparation exist because of gravity, and arise as a result of the density difference between particles and the metal matrix melt. Special experiments with a water-SiC suspension will be described on the next few pages in order to give a better understanding of particle sedimentation in MMCs synthesis. A variety of formula to estimate the viscosity of slurries will be presented. Finally, utilization of the phenomena will be discussed.

1.   Settling of SiC Particles in Water Suspension

In the discussed experiments, a well-mixed water-SiC suspension was poured into a transparent cup of d=120 mm diameter. The free surface of suspension reached a height of H=117 mm from the bottom of the cup (Fig. A1). A small quantity of suspension was taken out by sucking probe from different points of the liquid at different moments in time. The volume fraction of reinforcement particles in these small quantities of suspension was measured. The nominal diameter of particles was 20 microns. Experiments were carried out with initial particle concentration $V_f(0) = 10$ vol. %, 18 vol. % and 25 vol. %.

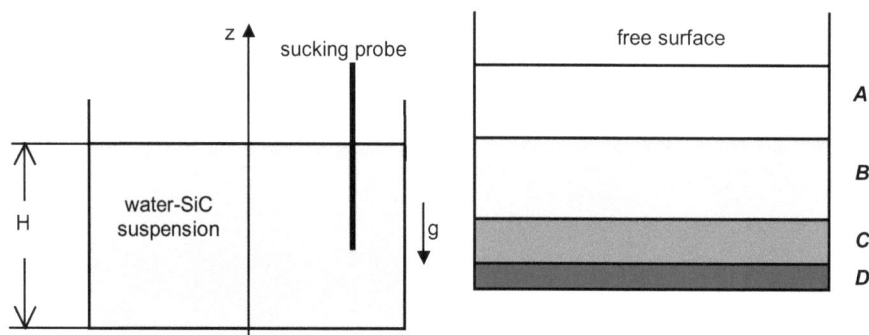

Fig. A1.   Schematic presentation of the   Fig. A2.   Dynamic of sedimentation:
experimental unit.   *A* - free region; *B* - region of
                                                      initial distribution; *C* - region of
                                                      increasing   distribution;   *D* -
                                                      region of final concentration

During sedimentation four regions of different particle density can be observed (Fig. A2). The region *A* is particle free, and in the adjacent region *B* particle concentration is equal to the initial concentration. The region *D*, formed at the bottom of cup, contains suspension with final (maximal) particle density. An intermediate region *C* with graded concentration is also formed. All the regions are present during the process. The width of the four zones changes with time. The settling starts when region *B* occupies the whole volume, but only *A* and *D* are present in the final state.

The results obtained are shown in Figs. A3-A11. Figs. A3-A5 relate to measurements of particle volume fraction at the levels 110 mm, 92 mm and 73 mm. Here the initial concentration of particles, $V_f(0)$, is 10 vol. %. Figs. A6-A8 depict the results for the same levels in the case $V_f(0)$ = 18 vol. %. Fig. A9 and Fig. A10 show measured values of particles concentration at levels 92 mm and 66 mm and Fig. A11 shows the movement of the upper boundary of the particlerich zone (the dividing boundary between regions *A* and *B* in Fig. A2) as functions of time in the case $V_f(0)$ = 25%.

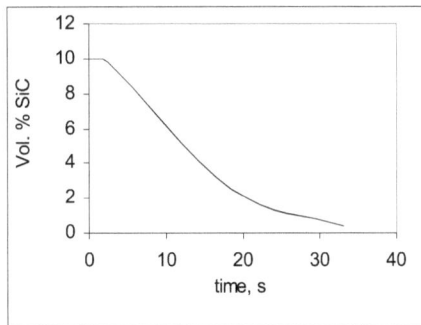

Fig. A3.   Dynamics of SiC volume fraction at H = 110 mm. Initial concentration $V_f(0)$ = 10 vol.%

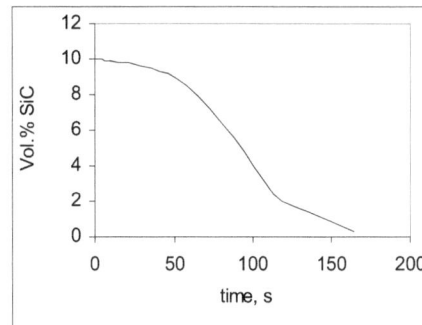

Fig. A4. Dynamics of SiC volume fraction at H = 92 mm. Initial concentration $V_f(0)$ = 10 vol.%

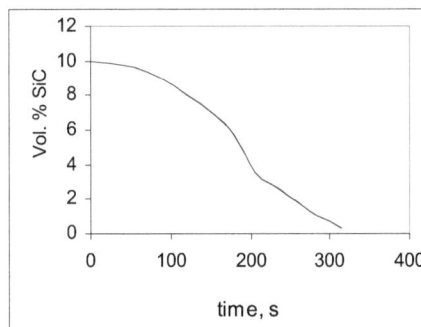

Fig. A5.   Dynamics of SiC volume fraction at H = 73 mm. Initial concentration $V_f(0)$ = 10 vol.%

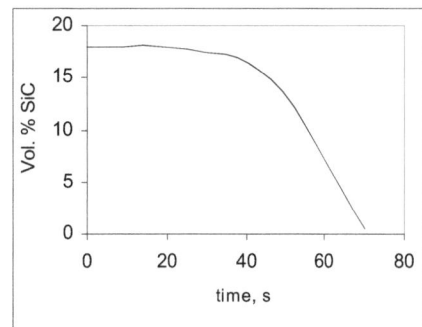

Fig. A6. Dynamics of SiC volume fraction at H = 110 mm. Initial concentration $V_f(0)$ = 18 vol.%

The curves in all the figures shown the dynamics of particle concentration V_f have a relatively slow decrease in V_f, but not as sharp as could be expected. The main reason for this is that the particles in the suspension have different diameters. According to image analysis results, the size distribution for used SiC particles of nominal diameter 20 μm is given in Fig. A12. Our measurements show that approximately 20 % of the particles have diameters greater then 22 μm, and less that 5 % have diameters smaller then 18 μm. It means that more then 20 % of the particles will settle at a velocity greater then the nominal velocity, which is associated with particles of the nominal diameter. On the other hand, when particles concentration at a certain point becomes smaller then 5 %, one can consider that the concentration is equal to zero for particles of nominal diameter.

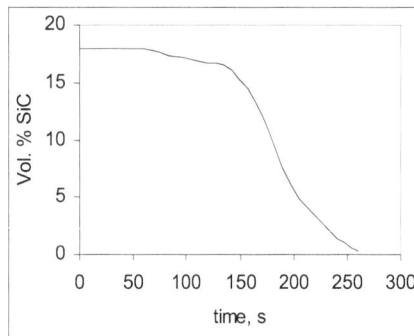

Fig. A7. Dynamics of SiC volume fraction at H = 92 mm. Initial concentration is $V_f(0) = 18$ vol.%

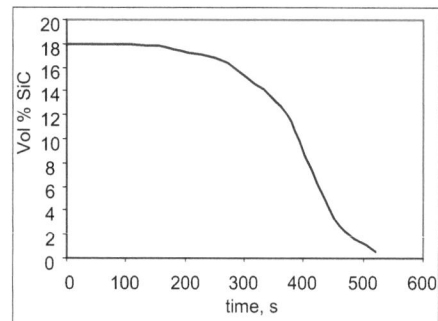

Fig. A8. Dynamics of SiC volume fraction at H = 73 mm. Initial concentration is $V_f(0) = 18$ vol.%

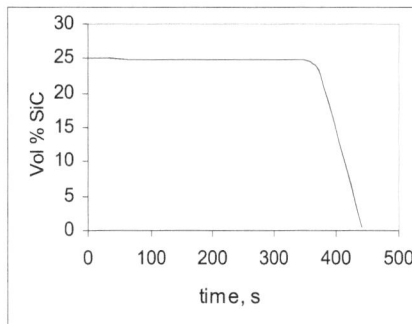

Fig. A9. Dynamics of SiC volume fraction at H = 92 mm. Initial concentration is $V_f(0) = 25$ vol.%

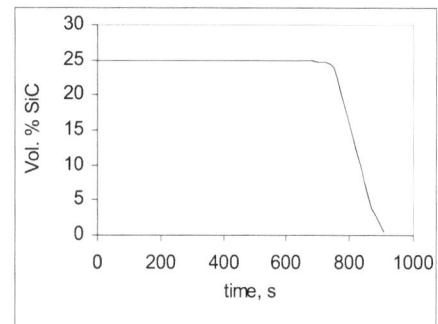

Fig. A10.   Dynamics of SiC volume fraction at H = 66 mm. Initial concentration is $V_f(0) = 25$ vol.%

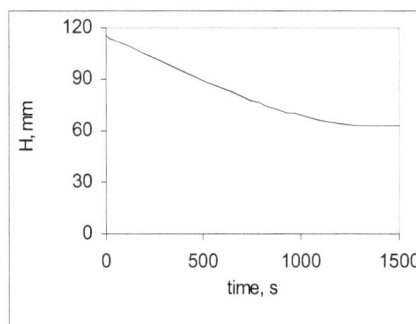

Fig. A11. Position of upper boundary of

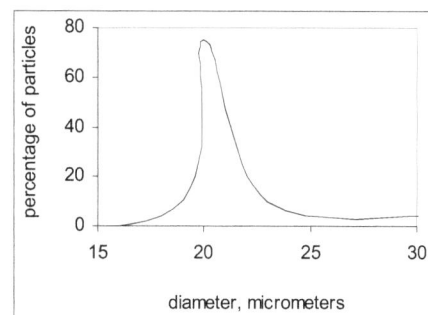

Fig. A12.   Percentage distribution of

particle rich zone as a function   used SiC particles of nominal
of time, $V_f(0) = 25$ vol.%   diameter 20 µm

2.   Theoretical Aspects

One of the forces determining the movement of particles in a liquid metal matrix during composite synthesis is the drag force, often called Stokes' force. This force is closely connected with a particular characteristic of real fluid, namely viscosity. Let us consider laminar fluid flow on a flat solid, Fig. A13. It is well known that viscosity η_0 is a local function and is defined as the coefficient proportionality coefficient between the modulus of the velocity gradient and the force f_F of internal friction in liquid per unit surface area, i.e.

$$f_F = \eta_0 \frac{dv}{dz}$$

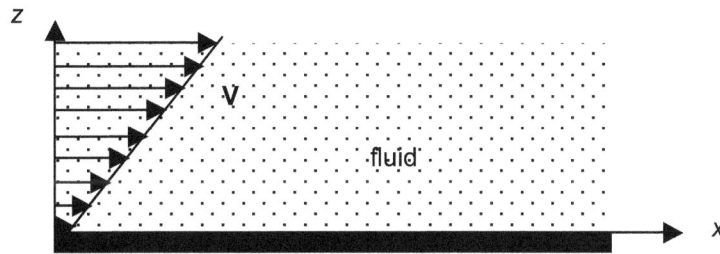

Fig. A13.   The field of velocity in laminar flat fluid flow

The last relation means that:

- the viscosity at a certain point in a liquid is equal to the friction force when the modulus of the velocity gradient $\frac{dv}{dz}$ at this point is equal to 1;

- the viscosity of a liquid reflects a fluid property which is connected only with its structure.

When a solid body moves in fluid, it experiences a drag force if the Reynolds number is small (laminar flow to be provided), the Stokes' law defines this drag force as:

$$f_D = A_S \eta_0 dv$$

Here η_0 is the viscosity of the liquid, d is the basic size of the solid, v is the relative velocity of the fluid with respect to the body and A_S is a shape factor. It is assumed that the distance between the body and the fluid boundary, for instance the walls of the vessel, is considerably greaterthan the size of the body. If the body is spherical, the basic size may be the radius r and the shape factor A_S equal to $6 \cdot \pi$. Hence, according to Stokes' law, the drag force acting on a sphere moving in liquid can be written as:

$$f_D = 6\pi \eta_0 r \, v \qquad\qquad (A1)$$

Eq. (A1) expresses Stokes' force applied to a moving particle in fluid when the following assumptions are valid:

1. the particle is spherical;
2. the Reynolds number is substantially less than 1 (laminar flow);
3. complete wettability (in other words, fluid velocity on the particle surface is equal to zero);

4. there is no solid boundary (i.e. other particles or walls of vessel) close to the particle.

When these conditions are not satisfied, the drag force has to be calculated by means of other relations. In the process of metal matrix composite synthesis, conditions 1, 3 and 4 are not valid and condition 2 has to be checked in every particular case, especially in centrifugal casting.

In the case that a single particle moves in liquid and all the above four conditions are valid, a crosssection of the velocity field looks like in Fig. A14. Otherwise, if one or more of the conditions are not valid, the velocity field around the particle changes its shape (Fig. A15) and use the nominal Stokes' law (A1) is not appreciate.

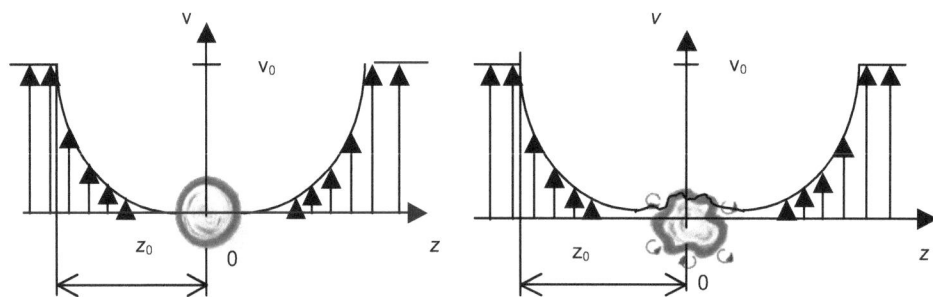

Fig. A14.   Schematic representation of velocity field around spherical particle

Fig. A15.   Schematic representation of velocity field around non-spherical particle

As was mentioned above, the drag force plays an essential role in the determination of particle movement. Stokes' law gives very simple expression for this force, and because of this it is applied in a large number of actual physical situations. In the case of composite synthesis, where a great number of particles move in liquid metal under gravity or centrifugal acceleration, the relation (A1) is used but in modified form. The modification has to compensate for the presence of many particles in the liquid and particles moving near to the vessel walls, saving the simplicity.

The presence of many particles affects the setting [5], and this process is referred to as hindered settling. This is basically due to upward fluid velocity increase during sedimentation with increase in concentration. This means that the relative velocity of the particles in respect to the liquid increases. We note that only this velocity is present in Stokes equation (A1).

Other physical reasons, which make the modification necessary are that the shape of particles is not spherical and that the particles interact with the vessel's walls. These lead to disturbance in the laminar fluid flow.

In order to evaluate the deviation of actual particle shape from spherical, a number Sf can be defined in the following way:

$$Sf = \frac{S}{V}$$

where S and V are the surface and volume of particle, respectively. For sphere with radius r

$$Sf = \frac{4\pi r^2}{4/3\pi r^3} = \frac{3}{r}$$

The effective radius of an arbitrary particle with volume V can be defined as follows:

$$r_{eff} = \sqrt[3]{\frac{3V}{4\pi}}$$

Then the number Sf becomes

$$Sf = \frac{S}{4/3\pi r_{eff}^3}$$

Values of Sf closer to $3/r_{eff}$ mean a shape closer to spherical, and conversely, a langer Sf means greater deviation from a sphere.

   The shape of actual particles is not spherical, and they can rotate with respect to their mass center during settling. This will violate laminarity of fluid flow and will result in a decrease in particle velocity. Generally, all real deviations from the Stokes' law requirements result in reduction of particle velocity, which means that the real drag force increases compared with the drag force evaluated by (A1).In order to have an integral description of multiparticle effects on a single particles movement and to retain the simplicity of the expression, an additional term η_V is used in Stokes law:

$$f_D = 6\pi\eta_0\eta_V r \, v \qquad\qquad \mu_V{\geq}1$$

   We call the complex $\eta=\eta_0\eta_V$ the adapted viscosity coefficient, and using it the law discussed here can be written in the form

$$f_D = 6\pi\eta r \, v \qquad\qquad\qquad (A2)$$

   Here $\eta_0=\eta_0\,(T)$ is the viscosity of pure metal matrix as a function of temperature T, and η_V expresses the effects of multiparticle movement, non spherical particle shape and the influence of the vessel's walls in an integral way. As was mentioned above the real viscosity η_0 characterizes the internal friction of the liquid and is related to its structure. The adapted viscosity coefficient characterizes the friction of the particle when it moves in a suspension, and this is related to the structure of the suspension, i.e. the nature of the liquid, shape and quantity of solid particles, and ratio between particle size dP and vessel diameter dD, i.e. dP/dD. There is no sense in considering adapted viscosity coefficient as measure of the inner friction of the suspension. Formally, the adapted viscosity plays the same role for Stokes' equation in the case of a liquid - solid particle suspension as the real viscosity for Stokes' equation in the case of liquid.

   Generally, the η_V function depends on the particles' volume fraction V_f, the shape factor Sf and the ratio dP/dD, i.e. $\eta_V = \eta_V(V_f, Sf, dP/dD)$. For a certain type of reinforcement particles Sf and dP/dD can be considered to be approximately constant and in this case $\eta_V = \eta_V(V_f)$. The only known relation for this function is $\eta_V(0) = 1$ and assuming continuity for η_V, it can be expressed as Taylor's series:

$$\eta_V\left(V_f\right) = \eta_V\left(0\right) + \eta_V'\left(0\right)V_f + \frac{\eta_V''\left(0\right)}{2!}V_f^2 + \frac{\eta_V'''\left(\xi\right)}{3!}V_f^3 \qquad \xi \in \left(0,V_f\right) \qquad \text{(A3)}$$

Using this implicit expression for $\eta_V(V_f)$ there is no need to have detailed mathematical model for very complicated physical reality of multiparticle movement in liquid. However, the coefficients $\eta_V'(0)$, $\dfrac{\eta_V''\left(0\right)}{2!}$ and $\dfrac{\eta_V'''\left(\xi\right)}{3!}$ in (A3) have to be estimated approximately on the basis of experimental data or other theoretical investigations.

An explicit mathematical model of single particle movement in conservative fields of gravity and centrifugal forces is discussed in [110]. According to this model particle sedimentation is described by a second order ordinary differential equation (the axis Z points in the opposite direction to gravitational acceleration, Fig. A1.):

$$\rho_1 V \frac{d^2 z}{dt^2} = -Vg\left(\rho_1 - \rho\right) - 6\pi\eta\varepsilon\frac{dz}{dt} \qquad \text{(A4)}$$

Here $V = 4/3\pi\varepsilon^3$, ρ_1, and ε are volume, density and radius of the particle; g and ρ are gravitational acceleration and the density of the liquid respectively. The first term on the right hand side is related to the buoyancy force which acts on the particle in the liquid, and the second term is related to the drag force according to Stokes' law. The coefficient η is the adapted viscosity coefficient. When the particle reaches its terminal velocity, the drag force applied to a particle settling in a viscous fluid becomes equal to the buoyancy force but with opposite direction, and the resulting force in (A4) becomes zero:

$$0 = -Vg\left(\rho_1 - \rho\right) - 6\pi\eta\varepsilon\frac{dz}{dt} \qquad \text{(A4')}$$

Because of this, the velocity remains constant until the end of settling. This transient period is relatively short in comparison with the duration of the whole sedimentation process. For this reason, in laminar flow regime, many authors use the solution of (A4') instead of that of (A4).

There are different approaches to the description of the multiparticle effect in sedimentation process. Richardson and Zaki [111] proposed a relationship between terminal setting velocity V_n and effective settling velocity of a suspension V_{eff}:

$$V_{eff} = V_n\left(1-C\right)^n$$

where C is the volume fraction of solid particles and n is a constant determined by the relation

$$n = 4.6 + 20 \frac{dP}{dD}$$

The most popular approach is to apply formula (A3) with special values on the coefficients. Einstein, based on a theoretical consideration [111], proposed for η_V thefollowing relation:

$$\eta_V = 1 + 2.5 V_f \qquad (A5)$$

Here, the last two terms in (A3) are neglected. It has been experimentally proved [111] that this relation is valid when $V_f < 0.02$.

In calculations for composite synthesis, the following relationship is most often [111]:

$$\eta_V = 1 + 2.5 V_f + 10.05 V_f^2 \qquad (A6)$$

Here, the last term in (A3) is neglected. Our experience shows that formula (A6) generates large errors in numerical simulations for the casting of real composite materials. The water model experiment described above proves that conclusion (see Table A1).

The expression (A3) can be written in the form

$$\eta_V = 1 + \alpha V_f + \beta V_f^2 + \gamma V_f^3 \qquad (A7)$$

The last formula for η_V turns into Einstein's formula (A5) for $\alpha = 2.5$, $\beta = 0$, $\gamma = 0$. and into the wellknown (A6) when $\alpha = 2.5$, $\beta = 10.05$ and $\gamma = 0$. In order to find more adequate expression of the adapted viscosity, the coefficients α, β and γ in the function η_V were determined using the experimental results, described above. More details about the procedure can be found in [3].

Taking into account the experimental results in Fig. A11, it was calculated that sedimentation of particles can occur up to a maximum volume fraction equal to 46,5 %, i.e. $V_{f, max} = 0.465$.

The following values for the coefficients are obtained: $\alpha = 18.5$, $\beta = 4.5$ and $\gamma = 170$, and formula (A7) looks like this:

$$\eta_V = 1 + 18.5 V_f + 4.5 V_f^2 + 170 V_f^3 \qquad (A8)$$

This formula is used in calculation of the thickness of the particle free zone as function of time in the experiments above. The same calculations are repeated applying formula (A6) and the results are compared in Table A1.

Using formula (A8) some additional calculations are performed. Fig. A16 presents particles distribution as a function of time. The above mentioned four regions are well shaped in the figure. For instance, at time 400 s, curve 3, the free region *A* covers distance between 53.8 mm and 117 mm. The region *B* is between 30.5 mm and 53.8 mm, and regions *C* and *D* are in the intervals 15.1-30.5 mm and 0.-15.1 mm respectively.

The dynamics of the regions *A*, *B*, *D* and *C* calculated by formulae (A6) and (A8) are shown in Fig. A17. The formula (A8), which provides more adequate description of the settling process (see Table A1) gives results that are very different in comparison with results obtained on the base of (A6).

Table **A1**. Size of the particle free zone during sedimentation of SiC in water.

Time, s		Size, mm	Size, mm	Size, mm
experimental	modified *	(measured)	(calculated by (6))	(calculated by (8))
For $V_f(0) = 0.10$				
35	32	6.4	11.5	5.
165	160	25.4	57.3	25.3
315	306	44.4	91.8	48.3
For $V_f(0) = 0.18$				
70	69	6.4	18.8	6.1
260	254	25.4	69.1	22.5
520	510	44.4	71.7	45.1
For $V_f(0) = 0.25$				
445	438	25.4	54.1	24.8
905	898	51.0	54.1	50.5

*The modification was performed due to the different particle sizes (see comments in paragraph II.) and according to curves on Figs.3-10

For most real situations we do not have complete wettability of solid particles melts, and their shape is not spherical. The formula (A8) provides higher accuracy in calculations for such cases.

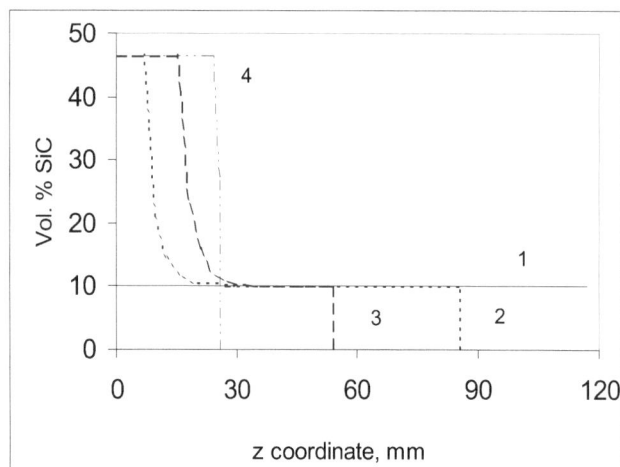

Fig. A16.   Calculated particle distribution during sedimentation. 1 = initial distribution; 2 = distribution after 200 s; 3 = distribution after 400 s; 4 = distribution at 600 s (final distribution)

As can be seen in Fig. A17a), there is a relatively long period of time (about 500-600 s) in which graded particle concentration exists in the suspension. Qualitatively the same can be obtained in case of flotation. This means that if one combines sedimentation and/or flotation with melt solidification, a great variety of graded structure can be obtained in particle reinforced MMCs. In certain melts, the velocity of particles movement can be manipulated by their size.

The direction of solidification and the solidification velocity can be controlled by the direction of heat removal, by the initial temperature of the composite slurry and the cooling rate. Gravity die casting and squeeze casting are two methods which are suitable for manufacturing of particle reinforced MMCs graded structure on the basis of sedimentation and/or flotation. Some possible structure conditions for these obtaining can be seen in Fig A18.

Another interesting fact which must be mentioned is that the time needed for complete sedimentation (or flotation) strongly depends on inclination angle of the settling tank, Fig. A19. This correlation is studied experimentally and described theoretically in [103]. The calculated velocity fields of the liquid and the sludge at time 300 s after the start of the settling process are shown in Fig. A20. The distribution of particle volume faction at time t = 100 s and t = 200 s for a set of inclination angles is given in Fig. A21. One can use this effect if a special arrangement of particle-free zone and graded particle-rich zone is required in a casting.

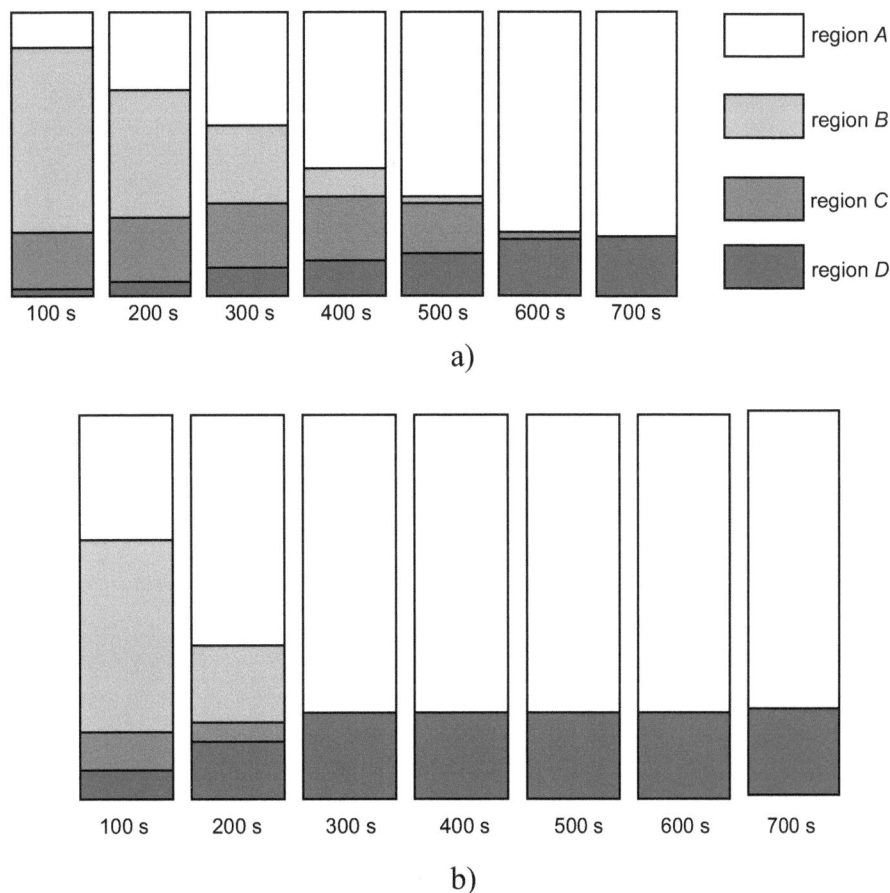

a)

b)

Fig. A17.   Schematic comparison of calculations performed for time-dependent distribution of SiC particles in water suspension of initial volume fraction $V_f(0) = 0.10$ SiC using a) formula (A8) and b) formula (A6).

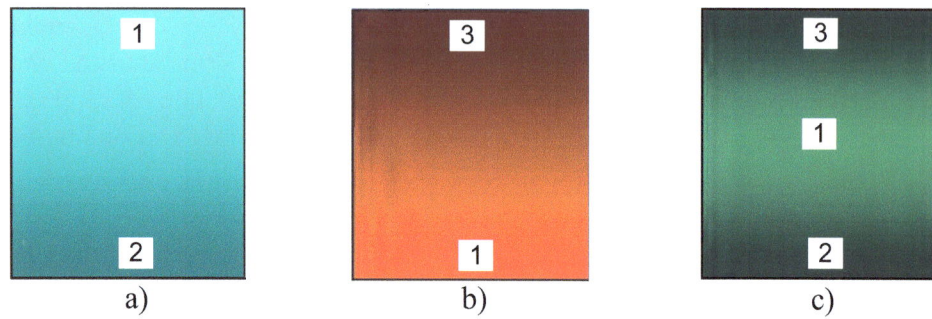

Fig. A18.   Schematic presentation of some graded structure in particle reinforced MMCs obtained by gravity or squeeze casting. 1 = particle free region; 2 = graded particlerich region, density of particles greater than melt density; 3 = graded particlerich region, density of particles lower than melt density. a) MMC with heavy particles, b) MMC with light particles and c) MMC with heavy and light particles.

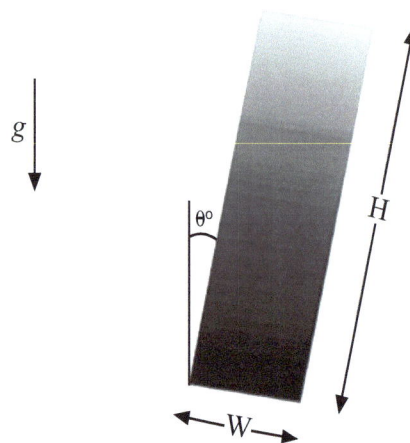

Fig. A19.   Inclination angle, $\theta°$, of a setting tank of height H and width W

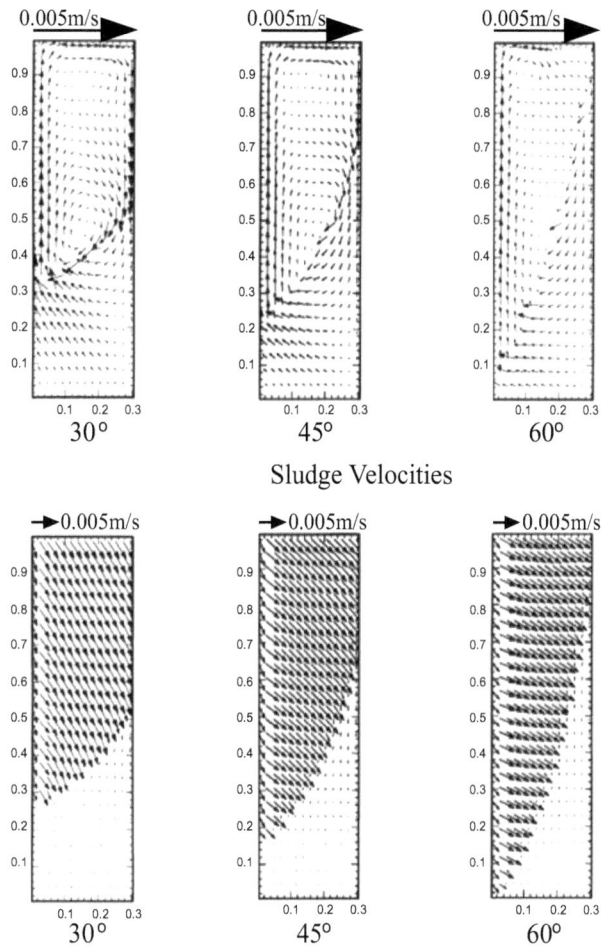

Fig. A20.   Velocity vectors at time t = 300 s as a function of inclination angle

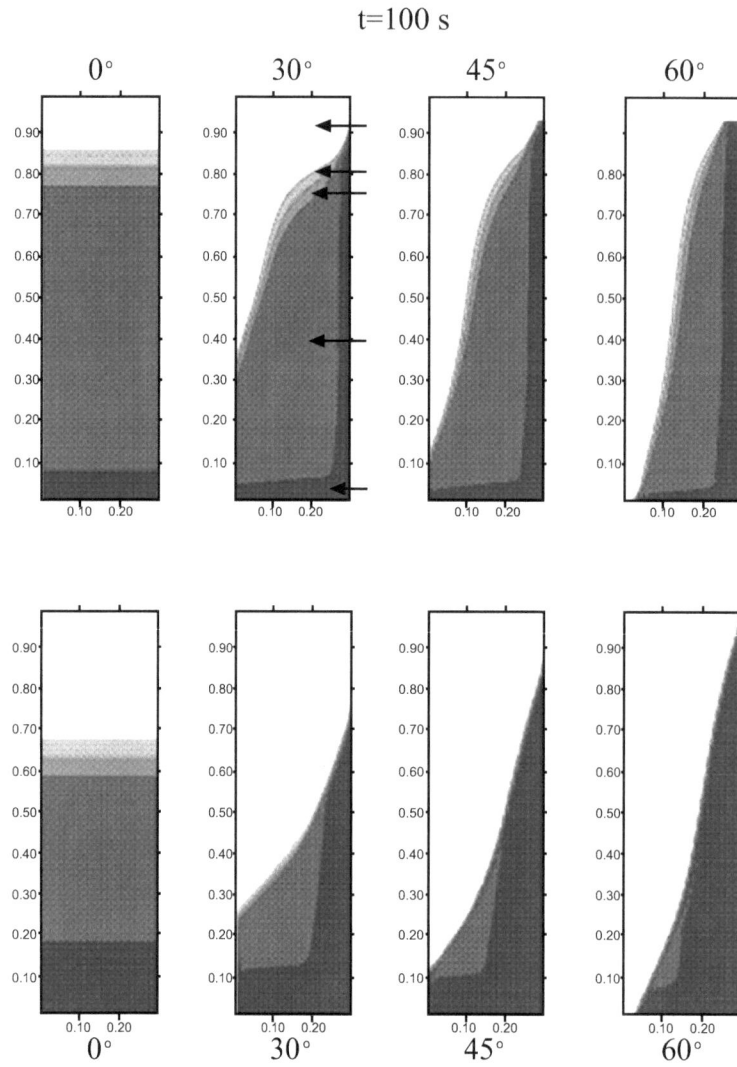

Fig. A21. Particle volume faction as a function of time t and inclination angle

Appendix B

GRADED STRUCTURES BY CENTRIFUGAL CASTING

Centrifugal casting offers greater potential for graded structure manufacture than gravity. The main reason for this is the higher pressure in the melt, which facilitates infiltration in a graded preform. Another reason is that usually the centrifugal force exceeds essentially the gravitational force, and because of this particles in composite slurry move faster than in the case of gravity casting. Moreover, by management of rotation speed one can control the magnitude of the centrifugal force, which means to control particle velocity in liquid composite during solidification. The latter is impossible in gravity casting and gives an important advantage of centrifugal casting for production of a variety of graded structures. A comprehensive mathematical model based on equations for both particle movement and solidification in rotating slurry, which would be capable of describing macrostructure formation in centrifugal casting of particle reinforced MMCs will be presented and short discussion on its application to the manufacture of graded structures will be analyzed.

1. Theoretical Analysis

Centrifugal casting of particlereinforced MMCs involves the solidification of a suspension of liquid metal and solid particles in a horizontally or vertically rotating mold. Segregation of solid particles dispersed in liquid rotating slurry occurs owing to centrifugal force. Particles are moved either to the outer or the inner part of the rotating mold because of their density difference with the melt. The process of particle segregation evolves together with the heat transfer process. These two processes define the changes in physical properties of the liquid metal, but the changes influence evolution of the processes. There are different approaches for describing the physical interactions which take place in this complex of phenomena characterizing centrifugal casting of particle-reinforced composites. Almost all models existing at present are one-dimensional, for simplicity because and of the fact that centrifugal acceleration is much greater than gravity. In real conditions, the two-dimensional heat transfer affects particle distribution.

Fluid Equilibrium in a Field of Conservative Forces

Let us consider a fluid in a cup rotated with constant angular rotational velocity ω. In this case the fluid will be in equilibrium in a gravity field. This means that the fluid does not move with respect to the cup in spite of rotation. At each point in the liquid, field of mass force $\mathbf{E}$ is applied, $\mathbf{E} = (E_x, E_y, E_z)$. For an arbitrary volume V, Fig. B1, with surrounding surface S and normal vector $\mathbf{n}$ on S, because of equilibrium, the following equation is valid:

$$\int_V \rho \mathbf{E} \, dv = \oint_S p \mathbf{n} \, ds \qquad (B1)$$

Jerzy Sobczak / Ludmil Drenchev (Eds.)

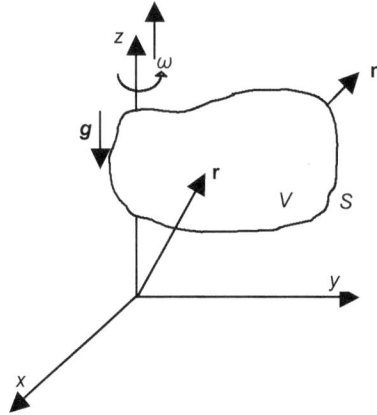

Fig. B1. Arbitrary volume of liquid in a rotating cup

Here ρ and p are density and pressure at point with radius-vector **r**. The left integral expresses the effective mass force distributed in the whole volume V, and the right integral gives the pressure resulting force. Applying the Gauss-Ostrogradski theorem

$$\oint_{S} p\mathbf{n}\,\mathrm{d}s = \int_{V} grad\,p\,\mathrm{d}v$$ (B2)

and making substitution, (B1) can be written as:

$$\int_{V} (grad\,p - \rho\mathrm{E})\,\mathrm{d}v = 0$$ (B3)

The last equation is valid for every arbitrary fluid volume V. This means that the function in brackets has to be zero at each point of the fluid, i.e.

$$grade\,p = \rho\mathbf{E}$$ (B4)

According to this equation, the force's components are:

$$, E_x = \frac{1}{\rho}\frac{\partial p}{\partial x}, \;\; E_y = \frac{1}{\rho}\frac{\partial p}{\partial y}, \;\; E_z = \frac{1}{\rho}\frac{\partial p}{\partial z}$$

Let us assume that the density depends only on pressure and define P as function of pressure as follows:

$$P(x,y,z) = \int \frac{1}{\rho}\mathrm{dp} + const$$ (B5)

It is evident that for mass force **E**, **E** = $grad$ P is valid. Whatever the function P, $grad\,P$ satisfies $rot\,grad$ P $\equiv$ 0. Therefore

$$rot\ \mathbf{E} = rot\ grad\ \mathrm{P} \equiv 0 \tag{B6}$$

This equation expresses the fact that fluid equilibrium is possible only in mass force fields for which $rot\ \mathbf{E} = 0$. Such fields are called conservative.

The fact that $\mathbf{E}$ is conservative force allows the introduction of a potential function U by relation:

$$\mathrm{E} = -\ grad\ \mathrm{U},\ i.e.\ \mathrm{E}_x = -\frac{\partial \mathrm{U}}{\partial x},\ \mathrm{E}_y = -\frac{\partial \mathrm{U}}{\partial y},\ \mathrm{E}_z = -\frac{\partial \mathrm{U}}{\partial z} \tag{B7}$$

An example of a conservative field is the gravity field. Let us consider that the z-axis of Cartesian coordinate system is directed upward and that acceleration due to gravity is a constant. In this case the field of gravity force can be defined as follows: $\mathrm{E}_{G\,x} = 0$, $\mathrm{E}_{G\,y} = 0$ and $\mathrm{E}_{G\,z} = -\ g$. Using the above equation we have

$$\mathrm{U}_G = g\ z + const \tag{B8}$$

If $\rho = const$, applying (B4), the following relation can be obtained:

$$\mathrm{p} = -\rho\ \mathrm{U}_G + const = -\rho\ g\ z + const \tag{B9}$$

In the case of constant angular velocity ω, the field of mass force has two components: gravity and rotational. If z axis is directed upward, the gravity mass force $\mathbf{E}_G = (\mathrm{E}_{G\,x}, \mathrm{E}_{G\,y}, \mathrm{E}_{G\,z})$ and its potential U_G remain as above. As is known, the acceleration on steady rotation is equal to the product of ω^2 and the distance between the axis of rotation and the considered point. For this reason, the rotational component of the mass force will be

$$\mathbf{E}_R = (\omega^2 x, \omega^2 y, \omega^2 z) \tag{B10}$$

and

$$\mathrm{U}_R = -\frac{1}{2}\ \omega^2 (x^2 + y^2 + z^2) + const \qquad (\mathbf{E}_R = -\ grad\ \mathrm{U}_R) \tag{B11}$$

For the case shown in Fig. B2, $\mathrm{U}_R = -\frac{1}{2}\ \omega^2 (x^2 + y^2) + const$ and the potential for the resulting mass force $\mathbf{E} = \mathbf{E}_G + \mathbf{E}_R$ will be the superposition of the two: $\mathrm{U} = \mathrm{U}_G + \mathrm{U}_R$. For this reason the following formulae are valid

$$\mathrm{U} = -\frac{1}{2}\ \omega^2 (x^2 + y^2) + g\ z + const, \tag{B12}$$

$$\mathbf{E} = -\ grad\ \mathrm{U} \tag{B13}$$

and

$$\mathrm{p} = \frac{\rho}{2}\ \omega^2 (x^2 + y^2) - \rho\ g\ z + const \tag{B14}$$

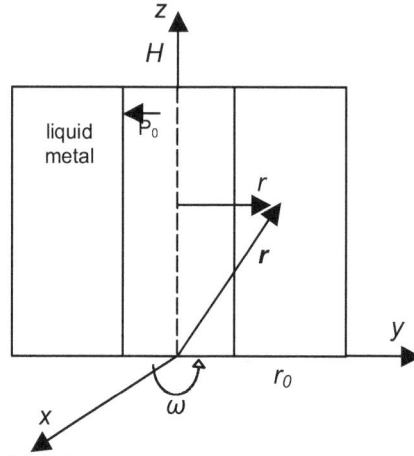

Fig. B2. Liquid metal in a high speed rotating cup

Formula (B14) gives the pressure distribution in fluid volume. For cylindrical symmetry, $x^2 + y^2 = r^2$ and

$$p = \frac{\rho}{2} \omega^2 r^2 - \rho \, g \, z + const \tag{B15}$$

In centrifugal casting, in the case of high rotation speed, liquid metal adjoins to the inner mold surface and forms a "tube", Fig. B2. On the basis of (B4) and (B13), using that dp = *grad*p·d*r* and   dU = *grad* U·d*r,* one can see that the next relation is valid:

$$dp = -\rho \, dU \tag{B16}$$

Applying (B12) and the relation $x^2 + y^2 = r^2$, the differential dU can be expressed like this:

$$dU = -\omega^2 r \, dr + g \, dz \tag{B17}$$

Thus relation (B16) becomes:

$$dp = \rho \, \omega^2 r \, dr - \rho \, g \, dz \tag{B18}$$

Let us integrate the last equation:

$$\int_{p_o}^{p} dp = \int_{r_o}^{r} \rho \omega^2 r \, dr - \int_{H}^{z} \rho g \, dz \tag{B19}$$

Assuming $\rho = const$, this relation gives

$$p = p_0 + \frac{\rho}{2} \omega^2 (r^2 - r_0^2) - \rho \, g \, (z - H) \tag{B20}$$

Here r_0 is the inner radius of the liquid and p_0 is the pressure on the inner surface of the liquid metal, for instance atmospheric pressure. This formula is obtained on the basis of conventional considerations but is very important. It expresses pressure in the liquid rotating metal as a function of the coordinates r and z, and it can be very useful if one models infiltration of molten metal in a porous preform or gas separation and gas pore formation during crystallization in a centrifugal casting. The pressure in the melt is proportional to the radius, which means that infiltration into a porous preform will be essentially easier at points far from the axis of rotation.

Particle Movement

Let us consider a spherical particle of radius ε that moves in rotating liquid with velocity **v**. A Cartesian coordinate system, attached to the rotating cup of liquid, is considered, Fig. B2. Axis z coincides with the rotation axis and is directed upward. The vector of angular velocity ω is directed along z-axes. For simplicity the particle is considered as material point with coordinate **x** = (x,y,z), coinciding with the mass center of the particle. First of all let us analyze the nature of the forces applied to the particle. The main force is the centrifugal force $\mathbf{F_R}$. Another is the force due to gravity $\mathbf{F_G}$. The drag force $\mathbf{F_S}$ owing to friction of the particle in the liquid is directed opposite to the velocity. The liquid, because of the density difference, exerts buoyancy force $\mathbf{F_B}$. The Coriolis force $\mathbf{F_C}$, which appears in respect to the rotating coordinate system, is perpendicular to the velocity. Applying Newton's law for particle of mass m, the following relation can be written:

$$m \frac{d}{dt} \mathbf{v} = \mathbf{F} = \mathbf{F_R} + \mathbf{F_G} + \mathbf{F_S} + \mathbf{F_B} + \mathbf{F_C} \tag{B21}$$

The first two terms on the right-hand side of the last equation are associated with conservative field, and according to (B12) and (B13)

$$\mathbf{F_R} + \mathbf{F_G} = m \, \mathbf{E} = - \, m \, grad \, U = - \, \rho_1 V \, grad \, [-\frac{1}{2} \omega^2 \, (x^2 + y^2) + g \, z + const \tag{B22}$$

or

$$\mathbf{F_R} + \mathbf{F_G} = (\rho_1 V \omega^2 x, \, \rho_1 V \omega^2 y, \, -\rho_1 V \, g) \tag{B23}$$

Here $V = \frac{4}{3} \pi \varepsilon^3$ is the volume of particle, and ρ_1 is its density.

The drag force $\mathbf{F_S}$, often called Stokes' force, can be defined like this:

$$\mathbf{F_S} = - \, 6 \, \pi \, \mu \, \varepsilon \, \mathbf{v} \tag{B24}$$

when the Reynolds number $Re = \frac{2v\varepsilon}{\mu} \rho$ is smaller than 1, and

$$\mathbf{F_S} = - \, 0.5 \, C_D \, \pi \varepsilon^2 \mu \, \mathbf{v} \, |\mathbf{v}| \tag{B25}$$

when Re>1, where C_D is the shape coefficient and μ is the viscosity of the liquid.

The buoyant force $\mathbf{F}_B$ is obtained from the next relation:

$$F_B = -\oint_S np\, ds = -\int_V grad\, p\, dv \tag{B26}$$

Because of the very small radius ε of particle, we can assume that *grad* p is constant in spherical volume of liquid with radius ε and center at point (x,y,z), and is equal to its value at this point. In this case, applying (B14), equation (B26) can be written as:

$$\mathbf{F}_B = -grad\, p \int_V dv = (-\rho V\omega^2 x,\ -\rho V\omega^2 y,\ \rho Vg) \tag{B27}$$

Equation (B27) is similar to (B23) but here the density of the liquid is involved and the sign is opposite. This means that buoyant force has opposite direction to the sum $\mathbf{F}_R + \mathbf{F}_G$.

The Coriolis force $\mathbf{F}_C$ is proportional to the vector product of $\mathbf{v}$ and $\boldsymbol{\omega}$

$$\mathbf{F}_C = 2\rho_1 V\, \mathbf{v} \times \boldsymbol{\omega} \tag{B28}$$

Here the vector product can be written as follows:

$$\mathbf{v} \times \boldsymbol{\omega} = \begin{vmatrix} i & j & k \\ v_x & v_y & v_z \\ \omega_x & \omega_y & \omega_z \end{vmatrix}, \; \mathbf{i, j, k} \text{ are unit vectors on the x, y and z, axes respectively.}$$

In the case considered $\boldsymbol{\omega} = (0, 0, \omega)$ and the vector product becomes

$$\mathbf{v} \times \boldsymbol{\omega} = \mathbf{i}\, \omega\, v_y - \mathbf{j}\, \omega\, v_x + \mathbf{k}\, 0 \tag{B29}$$

and the Coriolis force transforms to:

$$\mathbf{F}_C = (2\rho_1 V\omega v_y,\ -2\rho_1 V\omega v_x,\ 0) \tag{B30}$$

Having an explicit expression for all forces on the right side of (21), Newton's law gives the following system of three ordinary differential equations:

$$\rho_1 V\frac{d^2x}{dt^2} = V\omega^2 x(\rho_1 - \rho) - 6\pi\mu\varepsilon\frac{dx}{dt} + 2\rho_1 V\omega\frac{dy}{dt}$$

$$\rho_1 V\frac{d^2y}{dt^2} = V\omega^2 y(\rho_1 - \rho) - 6\pi\mu\varepsilon\frac{dy}{dt} - 2\rho_1 V\omega\frac{dx}{dt} \tag{B31}$$

$$\rho_1 V \frac{d^2 z}{dt^2} = -V g (\rho_1 - \rho) - 6\pi\mu\varepsilon \frac{dz}{dt}$$

The system (B31), together with the initial conditions

$$\mathbf{x}(0) = \mathbf{x}_0$$

and   (B32)

$$\mathbf{v}(0) = \mathbf{v}_0$$

has unique solution $\mathbf{x} = \mathbf{x}(t)$, which is the law of particle movement in a fluid rotating with constant velocity under the conditions discussed above. The coordinate system is steady attached to the rotating fluid.

The problem of the calculation of the function $\mathbf{x} = \mathbf{x}(t)$ which satisfies (B31) and (B32) can be solved by numerical methods.

Particular cases and modifications of the equations

The system (B31) has some basic features which can be very important and useful for particular cases, with practical application. First, the last equation in the system is independent of the others two. This means that the system can be divided into two independent parts, which can be solved separately.

Very often in the process of centrifugal casting of composite materials, because of high values of rotation angular velocity ω, the acceleration due to gravity is very small in comparison to the centrifugal acceleration, i.e.

$$g << \omega^2 x \qquad (B33)$$

In such cases, the third equation in (B31) can be ignored. Physically this means that there is no sedimentation or flotation. System (B31) can be reduced to

$$\rho_1 V \frac{d^2 x}{dt^2} = V \omega^2 x (\rho_1 - \rho) - 6\pi\mu\varepsilon \frac{dx}{dt} + 2\rho_1 V \omega \frac{dy}{dt}$$

$$\rho_1 V \frac{d^2 y}{dt^2} = V \omega^2 y (\rho_1 - \rho) - 6\pi\mu\varepsilon \frac{dy}{dt} - 2\rho_1 V \omega \frac{dx}{dt}$$
(B34)

Let us consider the following values of the parameters in equation (B31): $\varepsilon = 20 \cdot 10^{-6}$ m, $\mu = 0{,}15 \cdot 10^{-3}$ Pa·s, $\rho = 2390$ kg/m^3, $\rho_1 = 2700$ kg/m^3, $\omega = 2\pi\ 1800$ rad/min $= 2\pi\ 30$ rad/s.

The coefficient $2\rho_1 V\omega$ is approximately ten times smaller than the coefficient $6\pi\mu\varepsilon$. On the basis of this fact, one can disregard the Coriolis force. The calculations in this case give acceptable agreement with experimental results. If, for example, $\varepsilon = 40 \cdot 10^{-6}$ m, the above-mentioned proportion becomes 3, and if $\varepsilon = 70 \cdot 10^{-6}$ m the ratio is 1. Here the Coriolis force cannot be ignored. Fig. B3 depicts the tracks of different particles when Coriolis force was taken into account and when it is not. Fig. B4 shows the moving law of particles with different radii. When the rotation speed is high or the particle radius is large, the Coriolis force cannot be ignored.

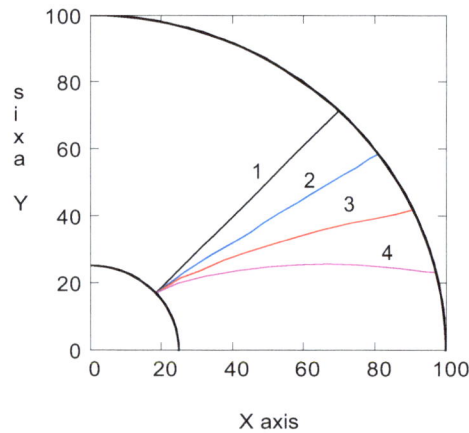

Fig. B3. Tracks of particle movement during centrifugal casting: 1 without considering Coriolis force; 2, 3, 4 considering the Coriolis force. 2 - ε = 100 μm, 3 - ε = 150 μm, 4 - ε = 190 μm

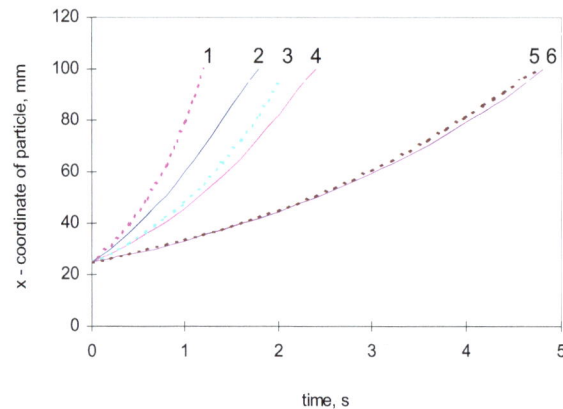

Fig. B4. Influence of Coriolis force on particle movement for different particle size:
1, 2 - ε = 200 μm; 3, 4 - ε = 150 μm;
5, 6 - ε = 100 μm. 1, 3, 5 – without considering the Coriolis force;
2, 4, 6 considering the Coriolis force.
Angular rotation speed 2π 1200 rad/min

In the case where the Coriolis force can be neglected, the equations in (B34) become identical. Only one equation has to be solved:

$$\rho_1 V \frac{d^2 r}{dt^2} = V \omega^2 r (\rho_1 - \rho) - 6 \pi \mu \varepsilon \frac{dr}{dt} \tag{B35}$$

Here r is the distance between axis of rotation and the particle. This equation can be modified to the following form:

$$\frac{d^2 r}{dt^2} + A\frac{dr}{dt} + B\,r = 0,$$

$$\text{where } A = \frac{9}{2}\frac{\mathrm{i}}{\tilde{n}_1 \mathring{a}^2} \qquad B = -\frac{\omega^2(\rho_1 - \rho)}{\rho_1}$$

(B36)

The problem of obtaining a function $r = r(t)$ which satisfies (B36) and the initial conditions $r(0) = r_0$ and $\frac{dr(0)}{dt} = v_0$ is one dimensional because gravity and the Coriolis forces are neglected. Here only radial movement takes place. This problem has a unique solution, which is written as:

$$(\text{i}) \qquad \text{if } \lambda^2 = A^2 - 4B > 0$$

$$r(t) = C_1\, exp[0.5\,(-A+\lambda)\,t] + C_2\, exp[0.5\,(-A-\lambda)\,t],$$

where $\qquad C_1 = \dfrac{r_0(\lambda + A) + 2v_0}{2\lambda}\quad$ and $\quad C_2 = \dfrac{r_0(\lambda - A) - 2v_0}{2\lambda}\,;$

$$(\text{ii}) \qquad \text{if } \lambda^2 = 4B - A^2 > 0$$

$$r(t) = C_1\, sin[0.5\,\lambda\,(t - C_2)]\, exp[-0.5A\,t],$$

where $\qquad C_1 = \sqrt{\dfrac{2}{\lambda}\left(v_0 + \dfrac{Ar_0}{2}\right) + r_0^2}\quad$ and $\quad C_2 = \dfrac{2}{\lambda}\,arcos\left(-\dfrac{r_0}{C_1}\right);$

$$(\text{iii}) \qquad \text{if } 4B = A^2$$

$$r(t) = (C_1\,t + C_2)\, exp[-0.5A\,t],$$

where $\qquad C_1 = v_0 + 0.5\,A\,r_0 \qquad$ and $\qquad C_2 = r_0$

Some authors cite as a solution of (B35) the expression:

$$r(t) = r_0\, exp\left[\frac{2\grave{u}^2(\rho_1 - \rho)\mathring{a}^2\,t}{9\,\mathrm{i}}\right]$$

(B37)

Obviously the above expression is not an exact solution because it does not satisfy (B35). This is an asymptotic solution of the equation considered for large values of t, and is an exact solution of

$$V\,\omega^2 r(\rho_1 - \rho) - 6\pi\mu\varepsilon\frac{dr}{dt} = 0$$

(B38)

This equation describes the situation where the particle moves in the following force equilibrium

$$\mathbf{F}_R + \mathbf{F}_B + \mathbf{F}_S = 0 \tag{B39}$$

which means

$$m \frac{d}{dt} \mathbf{v} = \mathbf{F}_R + \mathbf{F}_B + \mathbf{F}_S = 0 \tag{B40}$$

Calculation of Particle Segregation during the Centrifugal Process

The system of ordinary differential equations (B31) and initial conditions (B32) define the position of a single particle during the centrifugal process. When possible (see the discussion above), it can be reduced to an equation which has explicit solution according to the relation between its coefficients. But in a suspension of large particle volume fraction V_f, it is incorrect to use as the suspension viscosity the viscosity of pure liquid. There are a number of physical reasons to modify μ in (B24). In the case of multiparticle movement, instead of dynamic viscosity μ, an adapted viscosity coefficient μ_{eff} can be used that can be written in different forms (see Appendix A). The numerical calculations below are carried out with the formula

$$\mu_{eff} = \mu \left(1 + 18.5\ V_f + 4.5\ V_f^2 + 170\ V_f^3\right) \tag{B41}$$

in case of SiC particles, whose shape is closer to spherical and

$$\mu_{eff} = \mu \left(1 + 2.5\ V_f + 7.6\ V_f^2\right) \tag{B42}$$

in the case of graphite particles, whose shape is closer to that of flakes. More details can be found in [3].

Heat Conduction Model

Let us consider heat transfer between a cylindrical mold and cylindrical cast part in centrifugal casting. This system can be represented schematically as shown in Fig. B5.

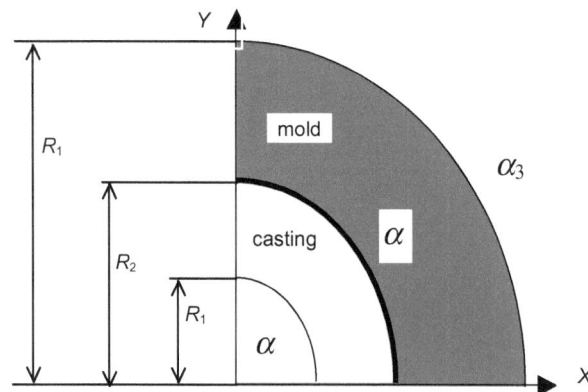

Fig. B5.   Schematic representation of cylindrical mold and casting.

R_1 = inner casting radius, R_2 = outer casting radius, R_3 = outer mold radius. α_1, α_2, α_3 - heat exchange coefficients on the inner casting surface, on interface between casting and mold, and on the molds outer surface, respectively.

The solidification of a hollow cylindrical cast part with height H, inner radius R_1 and outer radius R_2 is considered. For simplicity particle movement is not considered during filling. The initial mold temperature is T_0 and the pouring temperature is T_P. Constant heat exchange coefficients on the inner casting surface and outer mold surface are α_1 and α_3 respectively. The thermal interaction between casting and mold is defined by the coefficient α_2. At the beginning of the cooling process the coefficient α_2 is equal to h_0 and it decreases during solidification to h_F. To take into account shrinkage of the casting, the following relation for α_2 is used:

$$\alpha_2 = h_0 \left(\frac{h_F}{h_0} \right)^{\frac{d}{d_0}} \tag{B43}$$

where d is the current thickness of the solidified layer and d_0 is the casting thickness after solidification. The values of h_0 and h_F vary depending on the mold coatings.

The temperature field due to heat conduction in the mold is obtained as a solution of the equation:

$$C_m \frac{\partial T}{\partial t} = \lambda_m \left(\frac{\partial^2 T}{\partial r^2} + \frac{1}{r} \frac{\partial T}{\partial r} + \frac{\partial^2 T}{\partial z^2} \right), \quad t>0, r\in (R_2,R_3), z\in (0,H) \tag{B44}$$

and the temperature field in the casting is defined as a solution of the equation:

$$C_{eff}^c \frac{\partial T}{\partial t} = \lambda_c \left(\frac{\partial^2 T}{\partial r^2} + \frac{1}{r} \frac{\partial T}{\partial r} + \frac{\partial^2 T}{\partial z^2} \right), \quad t>0, r\in (R_1,R_2), z\in (0,H) \tag{B45}$$

where λ is thermal conductivity, $C_m = \rho_m c_m$, $C_{eff}^c = \rho_{eff} (c_{eff} +L_{eff} \frac{\partial f_L}{\partial T})$, c is the specific heat and L_{eff} is the latent heat of the melt. The simplest assumption, which is used here, is:

$$\frac{\partial f_L}{\partial T} = \frac{1}{\Delta T_{cr}}, \ \Delta T_{cr}= T_L - T_S \tag{B46}$$

Above, the index m indicates parameters related to the mold, upper index c to the casting, f_L is the relative mass fraction of liquid in the mushy zone and T_L and T_S are liquidus and solidus temperatures for the cast alloy respectively. The heat exchange coefficients on casting surface z=0 and the surface z=H are α_B and α_T respectively.

The thermal properties at each point of the composite are determined by the rule of mixture as a function of the particle volume fraction $V_f(t)$, which varies with time t:

$$\rho = [1- V_f(t)]\, \rho_L + \rho_p\, V_f(t)$$
$$c = [1- V_f(t)]\, c_L + c_p\, V_f(t)$$
$$\lambda = [1- V_f(t)]\, \lambda_L + \lambda_p\, V_f(t) \qquad (B47)$$
$$L_{eff} = [1- V_f(t)]\, L$$

The volume fraction depends on the particle concentration, speed of mold rotation, viscosity of the alloy and the difference in density between particle and alloy. A procedure for calculation of the volume fraction is discussed in [6]. Heat exchange at the inner casting surface, outer mold surface and casting-mold interface is considered to obey Newton's law. On the basis of the above assumptions typical boundary conditions for the heat transfer problem are defined.

2.   Numerical Experiments

Casting macrostructure formation consists of two basic processes, which run simultaneously: first, particle movement, modeled by (B31) or (B34), or (B35); and second, heat transfer and solidification, described by (B43-B47). In the applications discussed below these partially differential equations are solved numerically. The method of lines, which transforms the partial differential equations into a large system of ordinary differential equations, is applied. To find a solution of the obtained system of ordinary differential equation a standard software implementation of Adam's method is used. A variety of numerical simulations, which show the model features and explain many particular usual and unusual MMC structures obtained by centrifugal casting of Cu alloy/graphite, Al alloy/graphite and Al alloy/SiC$_P$ composites, are commented on in [9,10]. Here only two of them will be discussed. All numerical simulations are performed by a specially developed software product, which realize the mathematical model described above. In all figures produced by the software product, the dashed line on the left marks axis of rotation and the color scale relates to the concentration field in the cross section. Distribution of particles in the casting cross section is shown between the dashed line and the scale.

The graded structure obtained by numerical simulation for an Al alloy A356/graphite casting is shown in Fig. B6. Because of the lower density of graphite, the particles move to the inner casting region. It is easy to see the wellformed graded particle distribution.

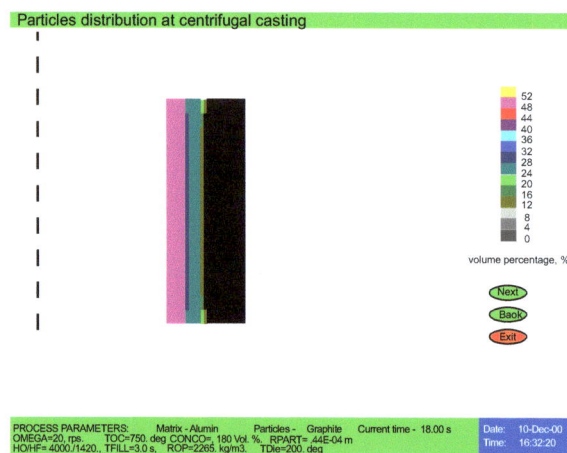

Fig. B6.   Graded structure in A356/graphite particle composite obtained under centrifugal casting. Particle size is 100 μm, rotation speed is 1000 rpm

A case of graded structure obtained in an actual Cu alloy C90300/graphite casting is given in Fig. B7 a). The density of alloy C90300 is approximately three times greater than the density of alloy A356. This casting has a free zone in its outer part, graded zone at the central region and a zone of maximal graphite content at the inner periphery. The technological parameters are as follows: rotation speed 1800 rpm, pouring temperature 1200°C, initial mold temperature 400°C, initial particle concentration 13 vol. %, an particle size 5 μm. Here, heat transfer parameters provided the solidification front moving faster than outer boundary of the particlerich region. For this reason an area with nonuniform distribution presents in the particlerich zone. The heat exchange coefficients between the casting and mold are estimated as $h_0 = 4000$ W/m^{2}s and $h_F = 1420$ W/m^{2}s. The dynamics of particle distribution during solidification can be seen in Figs. B7 b-d). Particles of 5 μm diameter are used as the reinforcing phase.

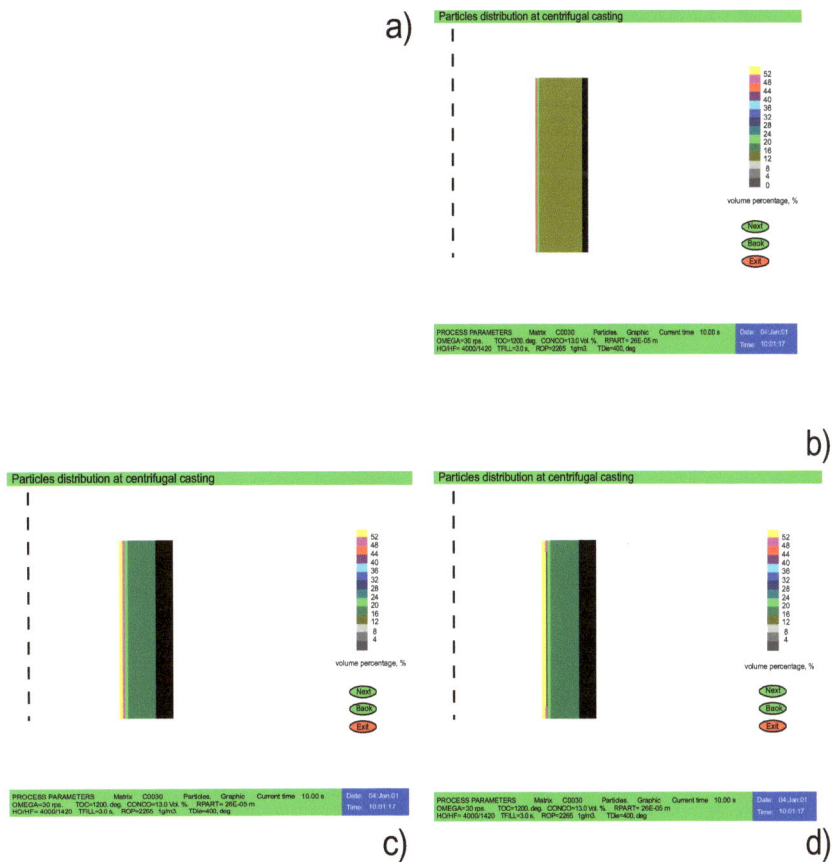

Fig. B7.   Centrifugally cast composite C90300/graphite system with graded area in the particle rich zone: a) cross section of the real casting; b), c) and d) particle distribution obtained by numerical simulation at times t = 10 s, t = 30 s and t = 50 s (final distribution), respectively.

Appendix C

LORENTZ FORCE IN LIQUID MEDIA

The force $\mathbf{f_L}$ which acts on a moving electrical charge q in magnetic field $\mathbf{B}$ is called the Lorentz force and is expressed as follows:

$$\mathbf{f_L} = q \cdot \mathbf{v} \times \mathbf{B} \qquad\qquad (C1)$$

where $\mathbf{v}$ is the velocity of the moving charge. The current in a conductor is the sum of the movements of all electrical charges that exist into the conductor, and here the force which acts on this conductor placed in the magnetic field will be a sum of the Lorentz forces acting on each charge. According to Ampere's low the force $\mathbf{f_A}$ that acts on a unit current element $\mathbf{j}$ in magnetic field is

$$\mathbf{f_A} = k \cdot \mathbf{j} \times \mathbf{B} \qquad\qquad (C2)$$

where k is a coefficient. Very often the force $\mathbf{f_A}$ is also called the Lorentz force.

Let us consider a liquid in which the electrical current density is $\mathbf{j}$ (Fig. C1), and which is displaced in magnetic field. The Lorentz force which acts on each unit volume in this liquid is defined by the relation

$$\mathbf{f_L} = \mathbf{j} \times \mathbf{B} \qquad\qquad (C3)$$

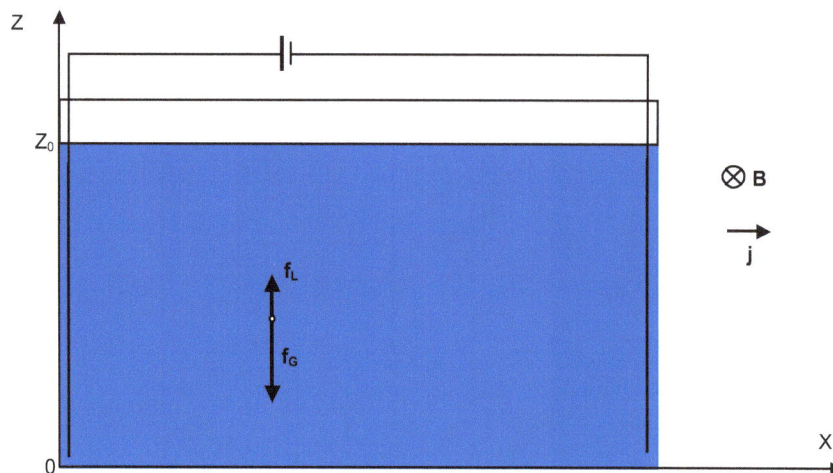

Fig. C1.   A vessel with liquid in which Lorentz force is applied

The last determines a field of Lorentz force at each point in the liquid. Formally, this field is similar to gravity field or centrifugal field. The direction of Lorenz force can be controlled by the direction of $\mathbf{j}$ and/or $\mathbf{B}$ and the combined effect of its simultaneous action with gravity, $\mathbf{f_G}$, in many cases looks like the "elimination" of gravity. For example, if the direction of Lorentz force is opposite to the gravity (Fig. C1) resulting force the applied at each point in the liquid will reduce the gravity or will even change its direction:

| $\mathbf{f_L}$ | < | $\mathbf{f_G}$ | - "reduction" of gravity;

| $\mathbf{f_L}$ | = | $\mathbf{f_G}$ | - "elimination" of gravity;

| $\mathbf{f_L}$ | > | $\mathbf{f_G}$ | - "change" of gravity direction.

Of course, the Lorentz force does not really affect gravity. These two forces exist together and act simultaneously. The combined effect when their sum is zero does not means that gravity is actually eliminated.

Let us consider single particle in a liquid and Lorentz force acting on the liquid (Fig. C1). Three different forces will be applied to the particle. The first is the buoyancy force (Archimedes force), which appears due to gravity and the difference in density of particle and liquid. For spheroidal particle this force is expressed as follows:

$$F_B = \frac{\pi d^3}{6}\left(\rho_2 - \rho_1\right)g \qquad (C4)$$

where ρ_1 is density of the liquid, ρ_2 is the particle density and d is the diameter of the particle. The second one is the drag (Stokes') force. This force is directed opposite to the velocity of the particle and it results from friction between the particle and the liquid. There are different formulae to define drag force. In the case of low particle velocity can be used

$$F_D = -\,3\,\pi\,\mu\,d\,v \qquad (C5)$$

Here μ is the viscosity of the liquid and v is the particle velocity.

The Lorentz force acts in different magnitude on the liquid and particles by reason of the different electrical conductivity of these two materials. As a result, a specific buoyancy force (similar to the Archimedes force) appears and causes particle movement up or down depending on the magnetic field direction. This force can be expressed by relation:

$$F_L = \pm\frac{3}{2}\,\frac{\sigma_1 - \sigma_2}{2\sigma_1 + \sigma_2}\,\frac{\pi d^3}{6}\,|\mathbf{j}\times\mathbf{B}| \qquad (C6)$$

Here σ_1 is the electrical conductivity of the liquid and σ_2 is the electrical conductivity of the particle.

Single particle movement in a liquid under Lorentz force is defined by the sum of above three expressions, i.e.

$$\frac{\pi d^3}{6}\rho_1\,\frac{d^2z}{d^2t} = F_B + F_D + F_L = \frac{\pi d^3}{6}(\rho_2 - \rho_1)\,g - 3\,\pi\,\mu\,d\,v \pm \frac{3}{2}\,\frac{\sigma_1 - \sigma_2}{2\sigma_1 + \sigma_2}\,\frac{\pi d^3}{6}\,|\mathbf{j}\times\mathbf{B}| \qquad (C7)$$

The initial conditions for (C7) can be defined as follows:

$$z(0) = z_0$$

$$\frac{\mathrm{d}z(0)}{\mathrm{d}t} = 0 \tag{C8}$$

The solution of the problem (C7,C8) defines completely the movement of a single particle in a liquid under a Lorentz force (Fig. C1). This is also valid in case of particle that does not conduct electricity, which means $\sigma_2 = 0$.

In the case of multiparticle movement (sedimentation or flotation) the viscosity of suspension is not equal to the viscosity of the liquid and must be modified as is discussed in [3]. The formation of a particle-rich region on the basis of the problem (C7,C8) can be described using numerical procedure described in detail in [3, 6].

The Lorentz force is used in many industrial applications. One of them is the removal of solid alumina from aluminum melts. The density of alumina is close to the density of liquid Al and because of this there is no sedimentation or flotation and it is difficult to purify the melts. If Lorentz force is applied to the melt, a buoyant force starts acting on the alumina particles according to (C6), and the particles float upward. It is easy a melt by removing solid particles after flotation.

Another very important application of Lorentz force is in the production to purify of graded structure in particle-reinforced composites. By means of current density and magnetic field (manipulating their direction and magnitude) a great variety of FGMs can be produced.

Appendix D

A COMPLEX MODEL OF GASAR STRUCTURE FORMATION

Physical processes that govern gasar structure formation include: heat transfer, gas diffusion in melt and solid, gas phase nucleation, and pore evolution. Heat transfer in the ingot (liquid and solid part) defines the solidification velocity, which in turn determinates gas concentration on the solid/liquid interface. This velocity is a component that also determinates the roughness of the interface. Greater roughness means more sites for gas phase nucleation. Gas diffusion in the liquid metal defines gas concentration ahead of the solidification front. Gas flux from the melt into the pores and their size are directly conditioned by gas concentration in the melt. Pore size and pore density (number of pores per unit area in a transversal cross section) depends on nuclei size, which is a function of surface tension σ_{lg}, and the number of nucleation sites also depends on quantity of impurity in the melt. All these processes are interconnected and determinate final structure in a complicated way.

1. Heat Transfer

We consider finite 3-D control volume that solidifies because of cooling from the bottom (open mold casting method). A gas mixture of argon and active gas (hydrogen or nitrogen) is above the melt. A cartesian coordinate system is defined as shown in Fig. D1. The temperature distribution, $T=T(x,y,z,t)$, in the solid and liquid sections is obtained as the solution of the following problem:

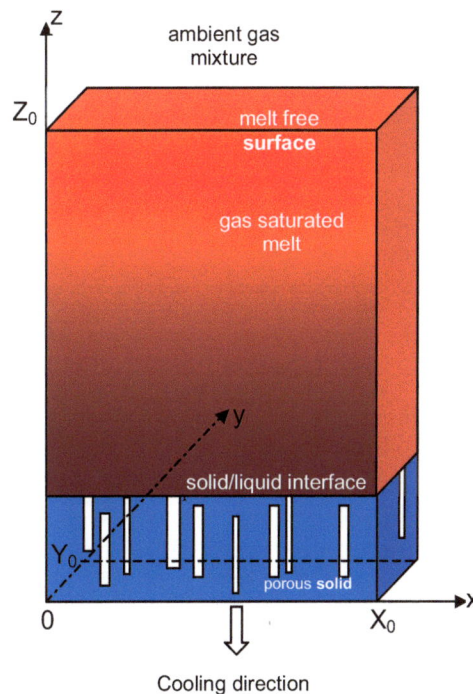

Fig. D1.   Schematic representation of the control volume

$$C_{eff} \frac{\partial T}{\partial t} = \frac{\partial}{\partial x}(\ddot{e} \frac{\partial T}{\partial x}) + \frac{\partial}{\partial y}(\ddot{e} \frac{\partial T}{\partial y}) + \frac{\partial}{\partial z}(\ddot{e} \frac{\partial T}{\partial z}) \tag{D1}$$

$$t>0, \ x \in (0, X_0), \ y \in (0, Y_0), \ z \in (0, Z_0)$$

where $C_{eff} = \rho (c + L_{eff} \frac{\partial f_L}{\partial T})$. Here $\rho = \rho(x,y,z,T)$, $c=c(x,y,z,T)$ and $\lambda=\lambda(x,y,z,T)$ are the density, specific heat and thermal conductivity, respectively. L_{eff} is the latent heat, and $f_L = f_L(T)$ is the relative liquid fraction in the two-phase (mushy) zone.

In many systems, especially when undercooling is small, the solid fraction may be assumed to be dependent on temperature only. It is assumed that L_{eff} is linear function between liquidus T_L and solidus T_S temperatures, and the derivative can be expressed as:

$$\frac{\partial f_L}{\partial T} = \begin{cases} 0 & \text{for } T < T_S \\ \dfrac{1}{T_L - T_S} & \text{for } T_S \le T \le T_L \\ 0 & \text{for } T > T_L \end{cases} \tag{D2}$$

The initial and boundary conditions are:

$$T(x,y,z,0) = T_0(x,y,z), \qquad x \in (0, X_0), \ y \in (0, Y_0), \ z \in (0, Z_0) \tag{D3}$$

$$\frac{\partial T(0,y,z,t)}{\partial x} = \frac{\partial T(X_0,y,z,t)}{\partial x} = 0, \qquad t>0, \quad y \in (0, Y_0), \ z \in (0, Z_0) \tag{D4}$$

$$\frac{\partial T(x,0,z,t)}{\partial y} = \frac{\partial T(x,Y_0,z,t)}{\partial y} = 0, \qquad t>0, \quad x \in (0, X_0), \ z \in (0, Z_0) \tag{D5}$$

$$-\lambda \frac{\partial T(x,y,0,t)}{\partial z} = \alpha_1 (T(x,y,0,t) - T_b), \qquad t>0, \quad x \in (0, X_0), \ y \in (0, Y_0) \tag{D6}$$

$$-\lambda \frac{\partial T(x,y,Z_0,t)}{\partial z} = \alpha_2 (T(x,y,Z_0,t) - T_c), \qquad t>0, \quad x \in (0, X_0), \ y \in (0, Y_0) \tag{D7}$$

where α_1 is the heat exchange coefficient between the ingot bottom and the mold, α_2 is heat exchange coefficient on the melts free surface, and T_c is gas temperature above the melt.

In the case of eutectic composition $T_L = T_S = T_{cr}$. In order to ensure numerical stability in solving (D1), it is considered that phase transformation takes place within a narrow temperature range assumed:

$$T_L = T_{cr} + 1 \text{ and } T_S = T_{cr} - 1$$

2.   Gas Diffusion

Gas diffusion in the melt occurs because of the non-uniform gas distribution formed during solidification and pore growth. Here only gas diffusion in liquid metal is considered and the 3-D diffusion problem is solved.

The dynamic of gas concentration in liquid C=C(x,y,z,t) are determined by solving of the diffusion equation:

$$\frac{\partial C}{\partial t} = D(\frac{\partial^2 C}{\partial x^2} + \frac{\partial^2 C}{\partial y_2} + \frac{\partial^2 C}{\partial z^2}), \qquad t>0,\ x\in(0,X_0),\ y\in(0,Y_0),\ z\in(Z_{S/L},Z_0) \qquad (D8)$$

with initial condition

$$C(x,y,z,0) = C_0 \qquad \text{where } C_0 = \eta(T)\sqrt{P_H} \qquad (D9)$$

and boundary conditions

$$\frac{\partial C(0,y,z,t)}{\partial x} = \frac{\partial C(X_0,y,z,t)}{\partial x} = 0 \qquad (D10)$$

$$\frac{\partial C(x,0,z,t)}{\partial y} = \frac{\partial C(x,Y_0,z,t)}{\partial y} = 0 \qquad (D11)$$

$$C(x,y,Z_0,t) = C_0 \qquad (D12)$$

$$C(x,y,Z_{S/L},t) = C_b \text{ for } (x,y) \text{ on pore/melt interface, where } C_b = \eta(T)\sqrt{P_b} \qquad (D13)$$

$$-D\frac{\partial C(x,y,Z_{S/L},t)}{\partial z} = C(x,y,Z_{S/L},t)\cdot(1-k)\cdot v_{cr} \text{ for } (x,y) \text{ on solid/melt interface} \qquad (D14)$$

In the equations above $Z_{S/L}$ is the coordinate of the solidification front, P_H is partial pressure of active gas above the melt, k is distribution coefficient for the metal-gas system considered, and v_{cr} is solidification velocity. P_b is the pressure in the bubble, which is defined as follow:

$$P_b = P_H + P_{Ar} + P_\sigma + P_g \qquad (D15)$$

Here P_{Ar} is the argon partial pressure in the gas mixture above the melt, P_g is the hydrostatic pressure on the solid/melt interface and P_σ is an extra pressure due to the curvature of gas/melt interface and can be expressed as follows:

$$P_\sigma = \frac{2\sigma_{gl}\cos(\theta_W - \theta_C)}{r_g} \qquad (D16)$$

Here r_g is the pore radius, θ_W and θ_C are contact angle of the melt on solid nucleation site and the half angle of the conical pit, respectively.

The function $\eta(T)$ expresses the temperature dependence of Sievert's law and in calculations the following expressions are used

$$\eta_1(T) = 0.722 \cdot 10^{-6} exp(-5234/T)$$
$$\eta_2(T) = 0.434 \cdot 10^{-6} exp(-5888/T)$$

(D17)

for liquid and solid copper, respectively.

The problems (D1-D7) and (D8-D16) are solved numerically using the finite difference method for space derivatives. A numerical, variable order version of the Adams-Gear's method is used to solve the obtained time depended system of ordinary differential equations. Let us mention that solidification velocity v_{cr} in (D14) is the link between the temperature and diffusion problems. This velocity in fact expresses the natural relation between solidification and gas transfer in the melt for the process discussed here.

3.   Random Processes in Structure Formation

Gas pore nucleation on solid/melt interface and simultaneous growth of solid and gas phases are the basic processes which determine the structure. Nucleation site and nucleus radius can be defined from the thermodynamic point of view in terms of the free energy changes associated with the process. It is well known that the probability of homogeneous bubble nucleation in a melt is negligible in comparison with the probability for heterogeneous nucleation under the same thermodynamic conditions. Heterogeneous nucleation appears at preferred nucleation sites, which may be microscopic pits and cracks on non-wetted inclusions in the melt or microscopic pits and cracks on the solidification front and mold surface. The critical radius, r_C, for nucleation is

$$r_C = \frac{2\sigma_{gl}}{\Delta g}$$

(D18)

and the activation barrier for nucleation is

$$\Delta G^* = \frac{16\pi \sigma_{gl}^3}{3\Delta g^2} S$$

(D19)

where Δg is the volume free energy change associated with gas phase nucleation and S is a shape factor, S=1 for a spherical nucleus and S <1 for other forms.

In the case of solid phase nucleation in pure metal melt, the volume free energy change is given by

$$\Delta g = \frac{L_V \Delta T}{T_m} \sim \Delta T$$

(D20)

Here ΔT is melt undercooling, L_V is latent heat and T_m is the melting point. By analogy, in the case of gas phase nucleation the volume free energy change can be expressed as follows:

$$\Delta g_C = \frac{M_C \Delta C}{C_0} \sim \Delta C \tag{D21}$$

where ΔC is melt supersaturation, M_C is the coefficient proportionality and C_0 is the equilibrium gas concentration in the melt as given by (D9). After substitution of (D21) in (D19) the activation barrier for gas phase nucleation can be expressed as:

$$\ddot{A}G_C^* = \frac{16\partial \acute{o}_{gl}^3 C_0}{3M^2 \ddot{A}C^2} S = \frac{A}{\ddot{A}C^2}, \quad \text{where} \quad A = \frac{16\partial \acute{o}_{gl}^3 C_0}{3M^2} S \tag{D22}$$

The nucleation rate can be expressed by the relation

$$N = N_S \frac{k_b T}{h} \exp\left(-\frac{\Delta G_m}{k_b T}\right) \exp\left(-\frac{\Delta G_C^*}{k_b T}\right) \tag{D23}$$

where N_S is the number of nucleation sites per unit surface of the solid/melt interface, ΔG_m is the activation energy for gas atom migration across the gas/melt interface, k_b is the Boltzmann constant, and h is the Planck constant.

The value of N_S depends on number of inclusions and roughness of the solid/melt interface. In all the numerical experiments discussed below this value is assumed to be constant. The value of ΔG_m is half of the activation energy for diffusion and is regarded as a constant, specific to each metal-gas system. ΔG_C^* depends on temperature and melt supersaturation. Here, only nucleation on the solidification front or close to it is considered, where the temperature can be assumed to be constant. Because of this, the relation (D22) is used in the form

$$\Delta G_C^* = \frac{A}{\Delta C^2} \tag{D24}$$

where A is constant.

The number of nuclei in the solidification front area Σ which appear from time t_1 to time t_2 can be calculated using the formula

$$I = \int_{t_1}^{t_2} \int_{\Sigma} N_S \frac{k_b T}{h} \exp\left(-\frac{\Delta G_m}{k_b T}\right) \exp\left(-\frac{\Delta G_C^*}{k_b T}\right) ds\,dt \tag{D25}$$

The probability of a nucleus to appearing at a point s on the solidification front is

$$P(s) = P_0 \exp\left(-\frac{\Delta G_C^*}{k_b T}\right) \tag{D26}$$

The coefficient P_0 does not depend on the position s.

The melt supersaturation at the solidification front depends on the position s and time. After substitution of (D24) in (D25) we obtain the number of nuclei in the form

$$I = \int\limits_{t_1}^{t_2}\int\limits_{\Sigma} N_S \frac{k_b T}{h} \exp\left(-\frac{\Delta G_m}{k_b T}\right) \exp\left(-\frac{A}{k_b T \cdot (C(x,y,Z_{S/L},t) - C_0)^2}\right) ds dt \qquad (D27)$$

where $C(x,y,Z_{S/L},t)$ is the gas concentration in the melt on solidification front obtained by solution of the problem (D8-D16). According to the discussion above all the quantities in (D27) can be assumed to be constant in an area Σ except $C(x,y,Z_{S/L},t)$ and C_0. Because of this (D27) is written in a simple form:

$$I = B_1 \int\limits_{t_1}^{t_2}\int\limits_{\Omega} \exp\left(-\frac{B_2}{(C(x,y,Z_{S/L},t) - C_0)^2}\right) ds dt \qquad (D28)$$

Here

$$B_1 = N_S \frac{k_b T}{h} \exp\left(-\frac{\Delta G_m}{k_b T}\right) \qquad (D29)$$

and

$$B_2 = -\frac{A}{k_b T} \qquad (D30)$$

are free parameters in the model, which are obtained by fitting the calculated results to the microstructures obtained in real experiments.

In the numerical experiments the number of nuclei on the solidification front from time t_1 to time t_2 is calculated by (D28). The probability $P(s)$ is estimated relatively for each cell of the generated mesh on solidification front, assuming $P_0=1$.

In the real case a gas nucleus may appear not only at point s_m of highest probability but also at a point s of probability $P(s)$ close to $P(s_m)$. For greater adequacy to the real situation, we consider that nuclei can arise at all positions (mesh cells) for which the probability $P(s)$ satisfies

$$qP(s_m) \le P(s) \le P(s_m) \qquad (D31)$$

We used q=0.7 in our calculations. All cells in which the probability satisfies (D31) are candidates for nucleation sites. When the number of these cells is greater than number I determined by (D28), the nucleation sites is randomly chosen among the candidates. If the number of candidates is equal to smaller than I, all cells are considered to be nucleation sites.

Nucleus size (initial pore radius), r_n, must be greater than the critical radius for nucleation r_C defined by (D18), $r_n > r_C$. The value of r_n depends on many parameters such as surface tension angle of pit or crack on inclusion surface or on the solidification front. This angle is different on different inclusion surfaces and at different sites on solidification front. Because of this all nuclei are of different size scattered around a mean value, $\overline{r_n}$. In the calculation discussed below, nucleation size is generated as random number in the range ($w_1 \overline{r_n}$, $w_2 \overline{r_n}$), i.e.

$$r_n \in (w_1 \overline{r_n} , w_2 \overline{r_n}), \qquad w_1 \le 1 \le w_2 \tag{D32}$$

The parameters w_1, w_2 and $\overline{r_n}$ is considered as input parameters. The effect of applying of (D31) and (D32) is that each time the numerical simulation is run without changing all the model parameters, the results obtained are similar but not the same. Obviously, this situation is closer to reality.

4.   Simultaneous Solid and gas Phase Growth

The software process that implements the considered model consists of three stages, which are repeated for each time step. In the first stage the combined problem (D1-D16) for temperature field and gas distribution is solved. In the second stage, on the basis of the solution for gas distribution on solidification front, $C(x,y,Z_{S/L},t)$, and relations (D26), (D28), (D31) and (D32), the number of nuclei and nuclei sites are determined. The third stage deals with the simultaneous growth of solid and gas phases. Boundary conditions (D13) and (D14) provide non-uniform gas concentration in the melt ahead of the solidification front, which will be the cause of gas fluxes in this area. At each time step, $\Delta t = t_2 - t_1$, the quantity of gas which passes trough the interface between a gas pore and the melt is estimated by the relation

$$Q = -\int_{t_1}^{t_2} \int_{S_{G/L}} D \frac{\partial C(x, y, Z_{S/L}, t)}{\partial z} ds dt \tag{D33}$$

Here $S_{G/L}$ is the interface between considered gas pore and the melt. Applying the relation

$$P_b V_p = \frac{Q}{\mu} RT_s \tag{D34}$$

the volume V_p of the porefor the time step being considered is found. Here P_b is the pressure defined by (D15) and μ is the molar mass of active gas. The volume V_p and ordered porosity ingot increment related to this time step determine the diameter of pore being considered.

To solve problem (D1-D16) a uniform mesh is defined in the initial (liquid) control volume, Fig. D1: Δx is the increment in the x direction, $L \cdot \Delta x = X_0$, Δy is the increment in the y direction, $M \cdot \Delta y = Y_0$, Δz is the increment in z – direction, $N \cdot \Delta z = Z_0$. Layer k is determined as parallel to x and y coordinate axes and occupies the volume between $z_{k-1} = (k-1)\Delta z$ and $z_k = k\Delta z$. After solidification of the layer k the height of this layer changes because some pores appear in it. We donate the new height $H_{op}(k)$. Other quantities used below are denoted as follows:

S_{xy} – area of transversal cross section of the control volume, i.e. cross section perpendicular to the z axis;

$S_p(k)$ – area of all pores in the cross section k;

V_{sol} – volume of solid fraction in layer k, $V_{sol} = S_{xy} \Delta z \rho_l/\rho_s$, ρ_l and ρ_s are the densities of melt and solid;

$V_p(i,k)$ – whole volume of pore i whose upper boundary is in layer k;

$d_p(i,k)$ – diameter of pore i in layer k;

$Q_{gas}(i,k)$ –   quantity (mol) of gas in pore i when its upper boundary is in layer k;

$S_{in}(i,k)$ –   area between pore i and melt in layer k;

$q_g(i,k)$ – gas flux in pore i in layer k;

$Q_{in}(i,k)$ –   quantity (*mol*) of gas coming into pore i when layer k solidifies;

Δt_k – time for solidification of layer k.

For estimation of $q_g(i,k)$ we use the expression

$$q_g(i,k) = - D \; grad \; C \tag{D35}$$

calculated for the element above pore i in layer k. The area of effective contact between pore i and melt in layer k is obtained as follows:

$$S_{in}(i,k) = \frac{\partial d_p^2(i,k)}{4} \tag{D36}$$

and $Q_{gas}(i,k)$ is expressed on:

$$Q_{gas}(i,k) = Q_{gas}(i,k\text{-}1) + Q_{in}(i,k) \tag{D37}$$

where $Q_{in}(i,k)$ is

$$Q_{in}(i,k) = q_g(i,k) \; S_{in}(i,k) \; \Delta t_k \tag{D38}$$

The well known relation for an ideal gas can be written for gas in a pore as:

$$P_b \left(V_p(i,k-1) - \frac{\pi d_p^2(i,k)}{4} H_{op}(k) \right) = Q_{gas}(i,k) \; R \; T_s \tag{D39}$$

Here P_b is defined by (D15) and R is the gas constant. After simple transformations an expression for $d_p(i,k)$ can be obtained

$$d_p(i,k) = 2 \sqrt{\frac{Q_{gas}(i,k) \, R \, T_s - P_b \, V_p(i,k-1)}{\pi \, P_b \, H_{op}(k)}} \tag{D40}$$

For $S_p(k)$ the following relation is valid

$$S_p(k) = \sum_i \frac{\pi \, d_p^2(i,k)}{4} \tag{D41}$$

Another relation that can be written is

$$S_p(k) \, H_{op}(k) + V_{sol} = S_{xy} \, H_{op}(k) \tag{D42}$$

Expressions D40-D42 are a system of algebraic equations which determine pore diameter, total area of pores and height of ordered porosity solid in relation to layer k. The solution of the system provides complete information for these three structure parameters.

5.   Numerical Experiments and Discussion

All the simulations discussed below are performed for copper/hydrogen system. Physical parameters for copper, which are used, can be found in Table D1.

The gasar ingot is obtained in a metal mold cooled predominantly at the bottom (open mold casting). The active gas is hydrogen and melt saturation is realized by gas mixture of hydrogen and argon. Initial melt temperature is 1473 K and solidification starts after complete melt saturation. Equilibrium gas concentration in liquid and solid copper is determined by relations (D9) and (D17).

Table **D1**. List of values of some basic characteristics copper used in the simulations.

Symbol	Quantity	Value	Units
D	Coefficient of diffusion for hydrogen in liquid (average) in solid (average)	$1.7 \, 10^{-8}$ $0.4 \, 10^{-8}$	m^2/s
λ	Thermal conductivity in liquid (average) in solid (average)	270 350	W/m^2K
ρ	Density liquid solid	8000 8400	kg/m^3
c	Specific heat for liquid (average) for solid (average)	470 515	J/kgK
L_v	Latent heat of fusion	205 000	J/kg
σ_{gl}	Surface tension	1.285	J/m^2

Thanks to the detailed and explicit description of the process of structure formation, many characteristics of structure and ingot and values of main physical fields can be obtained. The most important of these are:

-    temperature distribution in the melt and in the porous ingot;
-    local solidification velocity;
-    pore direction;
-    gas distribution in the melt;
-    local porosity;
-    average ingot porosity;
-    number of pores per unit area;
-    position of nucleus and closing of pores;

-    average pore diameter in transversal cross section.

The simulations can also be used for quantitative and qualitative estimation of relation between processing parameters and final structure. Some numerical experiments showing these possibilities are discussed below. More experiments carried out with this model and analyses can be found in [113].

Calculated gas concentration in the melt just ahead of the solidification front, in series, a time is shown in Fig. D2.

The results are related to the beginning of the structure formation, when the pore number in the volume is relatively small. In this case both pore number and gas concentration increase with time. Although the pore number increase, the inhomogenety in gas concentration also increase with time, compare Fig. D2 a) and Fig. D2 d).

Copper gasar, Par=0.7MPa, Ph=0.4MPa
Vsol~0.84E-03 m/s, 11 pores, time 6.72s

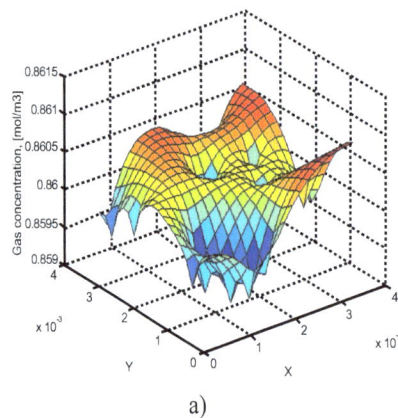

a)

Copper gasar, Par=0.7MPa,
Ph=0.4MPa Vsol~0.91E-03 m/s, 32
pores, time 8.86s

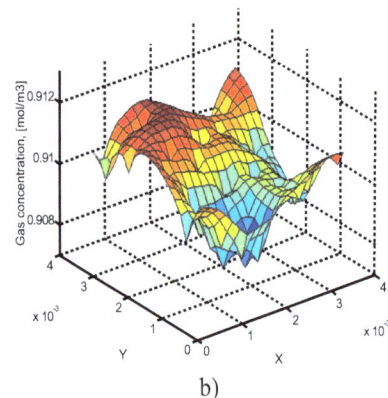

b)

Copper gasar, Par=0.7MPa, Ph=0.4MPa
Vsol~0.91E-03 m/s, 45 pores, time 11.00s

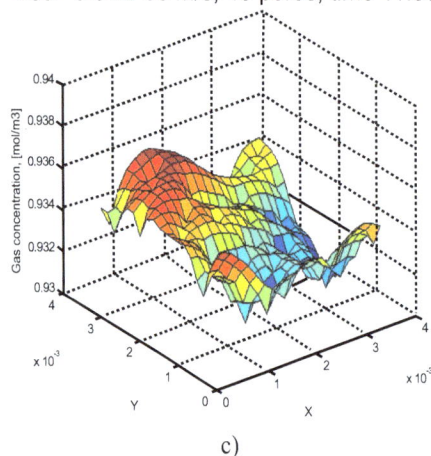

c)

Copper gasar, Par=0.7MPa,
Ph=0.4MPa Vsol~0.91E-03 m/s, 55
pores, time 13.18s

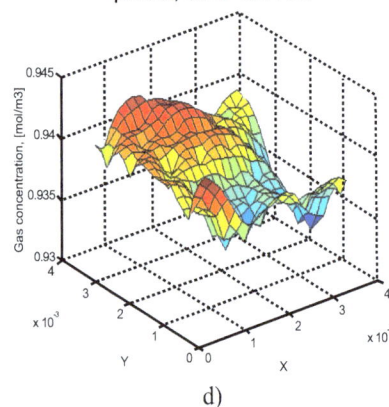

d)

Fig. D2.   Gas distribution in the melt on the solid/melt interface: a) at time 6.7s; b) at time 8.9s; c) at time 11.0s and d) at time 13.2s

The following experiment is intended to demonstrate influence of partial gas pressures on structure characteristics and the possibility of controlling ingot porosity through these pressures. The partial pressures in gas mixture above the melt are changed as shown in Fig. D3.

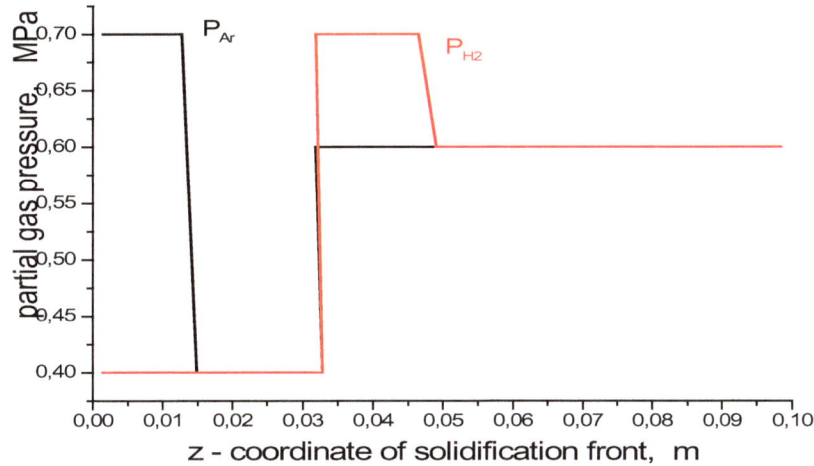

Fig. D3.   Changes of partial Ar and H_2 gas pressure versus position of solidification front

The rapid decreasing of partial argon pressure, P_{Ar}, when the solidification front passes through $z = 0.015$ m, causes a rapid increase in porosity and average pore diameter, Fig. D4 a). At the same time pore number per unit area and nucleus number per unit area also decrease rapidly.

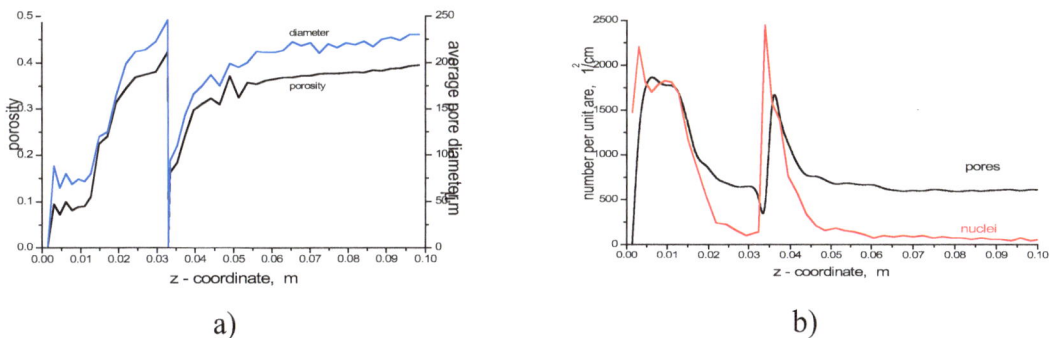

a)   b)

Fig. D4.   Change in structure caused by changes of partial gas pressures: a) porosity and average pore diameter at different cross-sections; b) pores per unit area and nuclei per unit area

The reduction in nuclei per unit area is greater than the reduction in pore number per unit area; Fig. D4 b). This is because the gas concentration at solid/melt interface is lower due to the increase in porosity. When the solidification front passes through $z = 0.033$ m both partial argon P_{Ar} and hydrogen P_{Ar} pressure increase rapidly, and porosity and average pore diameter decrease and reach zero. The very high gas pressure in the system causes reduction of gas volume in the pores, and all pores cease to exist. Then the whole quantity of gas rejected on the solid/melt interface remains in the melt ahead of solidification front and increases the probability of gas pore nucleation. After this many pores form and start to grow, as can be seen in Fig. D4 b). In this case again number of pore nuclei increases more rapidly than number of pores. This is because some nuclei do not become pores. They simply remain as very small gas bubbles in the solid. When solidification front passes through $z = 0.05$ m, P_{Ar} decreases from 0.7 MPa to 0.6 MPa. This reduction does not have sensible effect on the structure. The porosity and the average pore diameter slowly increase, because gas volume in the pores increases, Fig. D4a). At constant gas pressures nucleation and pores per unit area become approximately constant at relatively low level, Fig. D4 b). The local ingot porosity in the range 0.004 m < z < 0.015 m is about 0.09. From z = 0.015 up to z = 0.025 the local porosity rapidly increases up to 0.4. Then, because of the increase in partial pressures, the local porosity sharply decrease and a narrow non-porosity zone appeared in the ingot. At values of z greater than 0.06 m the local porosity reached approximately constant value of 0.37. This experiment demonstrates the possibility of manufacturing of ingot, whose local porosity varies over a wide range (in our case from 0 up to 0.4, see Fig. D4 a)) and is controllable by the gas pressures. Materials which graded porous structure and properties can be successfully produced by this technology.

REFERENCES

[1]   M. Koizumi, M. Niino, "Overview of FGM research in Japan", MRS Bull., vol. 20 (1), pp. 19-24, 1995.

[2]   A. Mortensen, S. Suresh, "Functionally graded materials and metal-ceramic composites, Part I: Processing", Int. Mater. Rev., vol. 40(6), pp. 239-265, 1995.

[3]   L. Drenchev, J. Sobczak, N. Sobczak, "Sedimentation Phenomenon and Viscosity of Water – SiC Suspension under Gravity Condition – a Water Model Study for Composites Synthesis", Colloids Surfaces A, vol. 197, pp. 203–211, 2002.

[4]   L. Yu, D. Gong, C. Wang, Z. Yang, L. Zhang, "Microstructure Analysis of W-Mo-Ti Functionally Graded Materials Fabricated by Co-sedimentation", Key Eng. Mater., vol. 249, pp. 299-302, 2003.

[5]   J. R. Gomes, A. R. Ribeiro, A. C. Vieira, A. S. Miranda, L. A. Rocha, "Wear Mechanisms in Functionally Graded Aluminium Matrix Composites: Effect of the Presence of an Aqueous Solution", Mater. Sci. Forum, vol. 492-493, pp. 33-38, 2005.

[6]   L. Drenchev, J. Sobczak, S. Malinov, W. Sha, "Numerical Simulation of Macrostructure Formation in Centrifugal Casting of Particle Reinforced Metal Matrix Composites. Part 1: Model Description", Modelling Simul. Mater. Sci. Eng., vol. 11, pp. 635–649, 2003.

[7]   A. Velhinho, P. Sequeira, R. Martins, *et al.*, "Evaluation of Al/SiC wetting characteristic in functionally graded metal-matrix composites by synchrotron radiation microtomography", Mater. Sci. Forum, vol. 423-425, pp. 263-268, 2003.

[8]   A. Velhinho, J. D. Botas, E. Ariza, J. R. Gomes, L. Rocha, "Tribocorrosion studies in centrifugally cast Al-matrix SiCp-reinforced functionally graded composites", Mater. Sci. Forum, vol. 455-456, pp. 871-875, 2004.

[9]   L. Drenchev, J. Sobczak, S. Malinov, W. Sha, "Numerical Simulation of Macrostructure Formation in Centrifugal Casting of Particle Reinforced Metal Matrix Composites. Part 2: Simulations and Practical Applications", Modelling Simul. Mater. Sci. Eng., vol. 11, pp. 651–674, 2003.

[10]   J. Sobczak, L. Drenchev, "Centrifugal Casting of Metal Matrix Composites", Transactions of the Foundry Research Institute, Special Issue No.1, Krakow, pp. 9-23, 2002.

[11]   C. Kang, P. Rohatgi, C. Narendranath, G. Cole, "Solidification analysis on centrifugal casting of metal matrix composites containing graphite particles", ISIJ Int., vol. 34, pp. 247–254, 1994.

[12]   L. Lajoye, M. Suery, Int. Symp. on Advances in Cast Reinforced Metal Composites, World Materials Congress (Chicago: ASM/TMS Metal Matrix Composites Committee), pp. 15–21, 1988.

[13]   S. Tsuru, N. Hayashi, T. Onoda, Y. Sakamoto, M. Hara, "Numerical simulation of a centrifugal process to fabricate permittivity graded FGM from Alumina/Epoxy mixture", Mater. Sci. Forum, vol. 492-493, pp. 459-464, 2005.

[14]   Y. Watanabe, S. Oike, I. Kim, "Formation of Compositional Gradient during Fabrication of FGMs by a Centrifugal *in-situ* Method", Mater. Sci. Forum, vol. 492-493, pp. 693-698, 2005.

[15]   P. D. Sequeira, Y. Watanabe, L. A. Rocha, "Particle distribution and orientation in Al-Al$_3$Zr and Al-Al$_3$Ti FGMs produced by the centrifugal method", Mater. Sci. Forum, vol. 492-493, pp. 609-614, 2005.

[16]   P. D. Sequeira, Y. Watanabe, L. A. Rocha, "Aluminum matrix texture and particle characterization in Al-Al$_3$Ti FGMs produced by a centrifugal solid-particle method", Solid State Phenomena, vol. 105, pp. 415-420, 2005.

[17]   F. Zhang, K. Trumble, K. Bowman, "Functionally graded Boron Carbide – Aluminum composites", Mater. Sci. Forum, vol. 423-425, pp. 73-76, 2003.

[18]   Y. Song, X. Mao, Q. Dong, L. Tang, Z. Ouyang, H. Liang, "Microstructure and properties of WCP reinforced ferrous gradient composites", Key Eng. Mater., vol. 336-338, pp. 2605-2608, 2007.

[19]   Y. Song, X. Mao, Q. Dong, B. Li, H. Liang, "Effects of the rotating speed of centrifugal machine on the gradient structure and properties of heavy cross-sectional WCP/Fe-C composites", Mater. Sci. Forum, vol. 475-479, pp. 1517-1520, 2005.

[20]   A. Oziębło, K. Konopka, E. Bobryk, M. Szafran, K. Kurzydłowski, "Al2O3-Fe functionally graded materials fabricated under magnetic field", Solid State Phenomena, 101-102, pp. 143-146, 2005.

[21]   A. Ozieblo, T. Wejrzanowski, K. Konopka, M. Szafran, K. Kurzydłowski, "Microstructure of Al_2O_3-Fe FGM obtained by modified slip-casting method", Mater. Sci. Forum, vol. 492-493, pp. 665-670, 2005.

[22]   C. Song, Z. Xu, X. Liu, G. Liang, J. Li, "In situ multi-layer functionally graded materials by Electromagnetic Separation method", Mater. Sci. Eng. A, vol. 393, pp. 164-169, 2005.

[23]   C. Song, Z. Xu, X. Liu, G. Liang, J. Li, "Study of in-situ Al/Mg2Si functionally graded materials by electromagnetic separation method", Mater. Sci. Eng. A, vol. 424, pp. 6-16, 2006.

[24]   C. Song, Z. Xu, X. Liu, J. Li, "In-situ Al/Al_3Ni functionally graded materials by electromagnetic separation method", Mater. Sci. Eng. A, vol. 445-446, pp. 148-154, 2007.

[25]   D. Kopeliovich, A. Shapiro, V. Shagal, US Patent No. US 6,273,970 B1, Aug. 14 2001.

[26]   Y. Kim, J. Choi, J. Park, K. Kim, E. Yoon, T. Nam, "The effect of electromagnetic vibration on the continuous elimination of inclusions in molten aluminum alloy by electromagnetic force", Mater. Sci. Forum, vol. 475-479, pp. 405-408, 2005.

[27]   M. Nygren, Z. Shen, "Spark plasma sintering: Possibilities and limitations", Key Eng. Mater., vol. 264-268, pp. 719-724, 2004.

[28]   A. Ohtsuka, A. Kawasaki, R. Watanabe, "Fabrication of Cu/Al2O3/Cu symmetrical functionally graded material by spark plasma sintering process", J. Jpn. Soc. Powder Metall., vol. 45, pp. 220-224, 1998.

[29]   Kikuchi K., Kang Y., Kawasaki A., "Optimization and fablication of Ni/Al2O3/Ni symmetric functionally graded materials", J. Jpn. Soc. Powder Metall., vol. 47, pp. 347-353, 2000.

[30]   S. Jin, H. Zhang, S. Jia, J. Li, "TiB2/AlN/Cu functionally graded materials (FGMs) fabricated by spark plasma sintering (SPS) method", Key Eng. Mater., vol. 280-283, pp. 1881-1885, 2005.

[31]   S. Jin, H. Zhang, S. Jia, J. Li, "TiB2/Cu electrode material fabricated via SPS", Mater. Sci. Forum, vol. 475-479, pp. 1555-1558, 2005.

[32]   K. Khor, K. Cheng, L. Yu, F. Boey, "Thermal conductivity and dielectric constant of spark plasma sintered aluminum nitride", Mater. Sci. Eng. A, vol. 347, pp. 300-308, 2003.

[33]   H. Zhang, J. Li, "Preparation of functionally graded Cu/AlN/Cu electrode materials for thermoelectric devices", Key Eng. Mater., vol. 336-338, pp. 2613-2615, 2007.

[34]   X. Tan, S. Qiu, W. He, D. Lei, "Functionally graded nano hardmetal materials made by spark plasma sintering technology", J. Metastable and Nanocrystalline Mat., vol. 23, pp. 179-182, 2005.

[35]   Z. Wang, M. L. Li., Q. Shen, L. M. Zhang, "Fabrication of Ti/Al2O3 Composites by Spark Plasma Sintering", Key Engineering Materials, vol. 249, pp. 137-140, 2003.

[36]   L. Ding, G. Luo, Q. Shen, L. Zhang, "Fabrication of Ti-Mg system composite with graded density at a low temperature by SPS method", Key Eng. Mater., vol. 249, pp. 291-294, 2003.

[37]   K. Zhang, Q. Shen, Q. Fang, Z. Wang, "Design and optimization of Al2TiO5/Al2O3 system functionally graded Mmaterials", Key Eng. Mater., vol. 249, pp. 141-144, 2003.

[38]   K. Nishiyabu, S. Matsuzaki, K. Okubo, M. Ishida, S. Tanaka, "Porous graded materials by stacked metal powder hot-press moulding", Mater. Sci. Forum, vol. 492-493, pp. 765-770, 2005.

[39]   E. Roncari, P. Pinasco., M. Nagliati, D. Sciti, A. Bellosi, "Tape casting of functionally graded AlN-SiC-MoSi2 composites", Key Eng. Mater., vol. 264-268, pp. 177-180, 2004.

[40]   Y. Kawakami, F. Tamai, T. Enjoji, K. Takashima, M. Otsu, "Preparation of tungsten carbide/stainless steel functionally graded materials by pulsed current sintering", Solid State Phenomena, vol. 127, pp. 179-184, 2007.

[41]   K. Tohgo, H. Araki, Y. Shimamura, "Evaluation of fracture toughness distribution in ceramic-metal functionally graded materials", Key Eng. Mater., vol. **345-346,** pp. 497-500, 2007.

[42]   K. Cho, I. Choi, I. Park, "Thermal properties and fracture behavior of compositionally graded Al-SiCp composite", Mater. Sci. Forum, vol. 449-452, pp. 621-625, 2004.

[43]   D. Song, Y. Park, Y. Park, I. Park I., K. Cho, "Thermomechanical properties of functionally graded Al-SiCp composites", Mater. Sci. Forum, vol. 534-536, pp. 1565-1568, 2007.

[44]   I. Zlotnikov, I. Gotman, L. Klinger, E. Gutmanas, "Combustion synthesis of dense functionally graded B_4C reinforced composites", Mater. Sci. Forum, vol. 492-493, pp. 685-691, 2005.

[45]   J. Du, Z. Zhou, S. Song, Z. Zhong, C. Ge, "Application of ceramics metal functionally graded materials on green automobiles", Mater. Sci. Forum, vol. 546-549, pp. 2283-2286, 2007.

[46]   H. Brinkman, J. Duszczyk, L. Kategerman, "Reactive hot pressing of functional graded metal matrix composites", Mater. Sci. Forum, vol. 308-311, pp. 140-145, 1999.

[47]   R. Jedamzik, A. Neubrand, J. Rödel, "Functionally graded materials by electrochemical processing and infiltration: Application to tungsten/copper composites", J. Mater. Sci., vol. 35, pp. 477-486, 2000.

[48]   Z. Ke, G. Chunb, "Powder metallurgy of tungsten alloy", Mater. Sci. Forum, vol. 534-536, pp. 1285-1288, 2007.

[49]   S. Novak, S. Beranic, "Densification of step-graded Al_2O_3-Al_2O_3/ZrO_2 composites", Mater. Sci. Forum, vol. 492-493, pp. 207-212, 2005.

[50]   J. Despois, A. Marmottant, Y. Conde, R. Goodall, *et al.*, "Microstructural tailoring of open-pore microcellular aluminium by replication processing", Mater. Sci. Forum, vol. 512, pp. 281-288, 2006.

[51]   W. Schaff, M. Hadenbruch, C. Korner, R. Singer, "Fabrication process for continous magnesium/carbon-fibre composites with graded fibre content", Mater. Sci. Forum, vol. 308- 311, pp. 71-76, 1999.

[52]   Jedamazik R., Neubrand A., Rodel J., "Characterization of electrochemically processed graded tungsten/copper composites", Mater. Sci. Forum, vol. vol. 308- 311, pp. 782-787, 1999.

[53]   D. Ilic, J. Fiscina, C. González-Oliver, F. Meucklich, "Properties of Cu-W functionally graded materials produced by segregation and infiltration", Mater. Sci. Forum, vol. 492-493, pp. 123-128, 2005.

[54]   A. Mattern, R. Oberacker, M. Hoffmann, "Tailored microstructures by multi-component pressure filtration", Key Eng. Mater., vol. 264-268, pp. 169-172, 2004.

[55]   A. Guntner, P. R. Sahm, "Graded metal matrix composites produced by multi-pouring method with controlled mold filling", Mater. Sci. Forum, vol. 308-311, pp.187-192, 1999.

[56]   C. Coddet, "Environmental protection of metal structures at high temperature: state of the art and future trends", Mater. Sci. Forum, vol. 461-464, pp. 193-212, 2004.

[57]   T. Jeon, H. Fang, Z. Lai, Z. Yin, "Development of functionally graded anti-oxidation coatings for carbon/carbon composites", Key Eng. Mater., vol. 280-283, pp. 1851-1856, 2005.

[58]   A. Yumoto, T. Yamamoto, F. Hiroki, I. Shiota, N. Niwa, "Fabrication of nanostructure composites in functionally graded coatings with supersonic free-jet PVD", Mater. Sci. Forum, vol. 492-493, pp. 341-345, 2005.

[59]   T. Tavsanoglu, O. Addemir, E. Basaran, S. Alkoy, "Processing and characterization of functionally graded Ti/TixCy/DLC thin film coatings", Key Eng. Mater., vol. 264-268, pp. 593-596, 2004.

[60]   O. Biest, L. Vandeperre, S. Put, G. Anné G., J. Vleugels, "Laminated and functionally graded ceramics by electrophoretic deposition", Key Eng. Mater., vol. 333, pp. 49-58, 2007.

[61]   S. Put, G. Anne, J. Vleugels, O. Biest, "Functionally graded ZrO_2-WC composites processed by electrophoretic deposition", Key Eng. Mater., vol. 206-213, pp. 189-192, 2002.

[62]   J. Tabellion, R. Clasen, "Manufacturing of advanced ceramic components via electrophoretic deposition", Key Eng. Mater., vol. 206-213, pp. 397-400, 2002.

[63]   E. Olevsky, X. Wang, "Graded powder composites by freeze drying, electrophoretic deposition and sintering", Mater. Sci. Forum, vol. 534-536, pp. 1533-1536, 2007.

[64]   C. Kunioshi, O. Correa, L. Ramanathan, "Gradient nickel – alumina composite coatings", Mater. Sci. Forum, vol. 530-531, pp. 261-268, 2006.

[65]   G. Anné, K. Vanmeensel, J. Vleugels, O. Biest, "Electrophoretic deposition as a novel near net shaping technique for functionally graded biomaterials", Key Eng. Mater., vol. 314, pp. 213-218, 2006.

[66]   T. Wei, A. Ruys, B. Milthorpe, "Hydroxyapatite-zirconia functionally graded bioceramics prepared by hot isostatic pressing", Key Eng. Mater., vol. 240-242, pp. 591-594, 2003.

[67]   T. Kokubo, "Novel inorganic materials for biomedical applications", Key Eng. Mater., vol. 240-242, pp. 523-528, 2003.

[68]   X. Miao, Y. Hu, J. Liu, B. Tio, P. Cheang, K. Khor, "Highly interconnected and functionally graded porous bioceramics", Key Eng. Mater., vol. 240-242, pp. 595-598, 2003.

[69]   S. Beranic, S. Novak, T. Kosmac, H. Richter, S. Mijic, " The preparation and properties of functionally graded alumina/zirconia-toughened alumina (ZTA) ceramics for biomedical applications", Key Eng. Mater., vol. 290, pp. 348-352, 2005.

[70]   T. Oberbach, C. Ortmann, S. Begand, W. Glien, "Investigations of an alumina ceramic with zirconia gradient for the application as load bearing implant for joint prostheses", Key Eng. Mater., vol. 309-311, pp. 1247-1250, 2006.

[70]   T. Oberbach, M. Christoph, W. Glien, "Development of an alumina ceramic with a gradient of concentration and microstructure for the application of load bearing implants", Key Eng. Mater., vol. 284-286, pp. 1027-1030, 2005.

[71]   J. Lackner, W. Waldhauser, L. Major, J. Morgiel et. al., in Foundation of Materials Design, Eds: K. Kurzydlowski, B. Major and P. Zieba, Research Signpost, Kerala, India, ISBN 81-308-0093-4, pp. 403-414, 2006.

[73]   V. I. Shapovalov, "Prospective applications of gas-eutectic porous materials (Gasars) in USA", Materials Science Forum, vol. **539-543,** pp. 1183-1188, 2007.

[74]   Ikeda T, Nakajima H, "Titanium coating of lotus-type porous stainless steel by vapor deposition technique', Mater. Lett., vol. 58, pp. 3807-3811, 2004.

[75]   T. Ikeda, M. Tsukamoto, H. Nakajima, "Fabrication of lotus-type porous stainless steel by unidirectional solidification under hydrogen atmosphere", Mater. Trans, vol. 43, pp. 2678-2684, 2002.

[76]   Nakajima H, Ikeda T, Hyun SK, "Fabrication of lotus-type porous metals and physical properties", in: Banhart J and Fleck NA (Eds.), Cellular Metals: Manufacture, Properties and Applications, MIT Verlag, Berlin, pp. 191-198, 2001.

[77]   T. Ikeda, T. Aoki, H. Nakajima, "Fabrication of lotus-type porous stainless steel by continuous zone melting technique and mechanical property", Metall. Mater. Trans. A, vol. 36, pp. 77-86, 2005.

[78]   H. Nakajima H, Hyun SK, Park JS, Tane M, "Fabrication of Lotus-type Porous Metals by Continuous Zone Melting and Continuous Casting Techniques", Materials Science Forum, vol. 539-543, pp. 187-192, 2007.

[79]   H. Nakajima, S. K. Hyun, M. Tane, T. Nakahata, "Fabrication and properties of porous materials with directional elongated pores", Materials Science Forum, vol. 512, pp. 295-300, 2006.

[80]   L. Drenchev, J. Sobczak, R. Ashtana, S. Malinov, "Mathematical modeling and numerical simulation of ordered porosity metal materials formation", J. Comp.-Aid. Mater. Des., vol. 10, pp.35-47, 2003.

[81]   L. Drenchev, J. Sobczak., S. Malinov, W. Sha, "Modelling of structural formation in ordered porosity metal materials", Modelling Simul. Mater. Sci. Eng., vol. 14, pp. 663-678, 2006.

[82]   A. Simone, L. Gibson, "The tensile strength of porous copper made by the GASAR process", Acta Mater., vol. 44, pp. 1437-1447, 1996.

[83]   H. Nakajima, S. Hyun, K. Ohashi, K. Ota, K. Murakami, "Fabrication of porous copper by unidirectional solidification under hydrogen and its properties", Colloids Surf. A, vol. 179, pp. 209-214, 2001.

[84]   K. Ota, K. Ohashi, H. Nakajima, "Internal friction in lotus-structured porous copper with hydrogen pores", Mater. Sci. Eng. A, vol. **341,** pp. 139-143, 2003.

[85]   Nakajima H, "Fabrication of Lotus-type Porous Metals, Intermetallic Compounds and Semiconductors", Materials Science Forum, vol. 502, pp. 367-371, 2005.

[86]   J. S. Park, S. K. Hyun, M. Tane, H. Nakajima, "Pore morphology of lotus-type porous copper fabricated by continuous casting technique", Solid State Phenomena, vol. 124-126, pp.1725-1730, 2007.

[87]   K. Rozniatowski, B. Ralph, K. Kurzidlowski, in Foundation of Materials Design, Eds: K. Kurzydlowski, B. Major and P. Zieba, Research Signpost, Kerala, India, ISBN 81-308-0093-4, pp. 175-194, 2006.

[88]   J. R. Gomes, A. S. Miranda, L. A. Rocha, R. F. Silva, "Effect of functionally graded properties on the tribological behaviour of aluminium matrix composites", Key Engineering Materials, vol. 230-232, pp. 271-274, 2002.

[89]   A. Velhinho, P. Sequeira, F. Braz, J. Botas, L. Rocha, "Al/SiCp functionally graded metal-matrix composites produced by centrifugal casting: Effect of particle grain size on reinforcement distribution", Mater. Sci. Forum, vol. 423-425, pp. 257-262, 2003.

[90]   J. Gomes, L. Rocha, S. Crnkovic, R. Silva, A. Miranda, "Friction and wear properties of functionally graded aluminum matrix composites", Mater. Sci. Forum, vol. 423-425, pp. 91-96, 2003.

[91]   J. Bonarski, in Foundation of Materials Design, Eds: K. Kurzydlowski, B. Major and P. Zieba, Research Signpost, Kerala, India, ISBN 81-308-0093-4, pp. 157-174, 2006.

[92]   H. Uzun, "Friction welding of functionally graded composites", Key Eng. Mater., vol. 264-268, pp. 659-662, 2004.

[93]   S. Schmauder, U. Weber, E. Soppa, "Computational micromechanics of heterogeneous materials", Key Eng. Mater., vol. 251-252, pp. 415-422, 2003.

[94]   C. Zhang, J. Sladek, V. Sladek, "Numerical analysis of cracked functionally graded materials", Key Eng. Mater., vol. 251-252, pp. 463-471, 2003.

[95]   M. Zhang, P. Zhai, Q. Zhang, "Effective thermal conductivity of functionally graded composite with arbitrary geometry of particulate", Key Eng. Mater., vol. 297-300, pp. 1522-1528, 2005.

[96]   O. Inan, S. Dag, F. Erdogan, "Three dimensional fracture analysis of FGM coatings", Mater. Sci. Forum, vol. 492-493, pp. 373-378, 2005.

[97]   P. Zhai, G. Chen, Q. Zhang, "Creep property of functionally graded materials", Mater. Sci. Forum, vol. 492-493, pp. 599-603, 2005.

[98]   J. Sladek, V. Sladek, C. Zhang, "Dynamic response of a crack in a functionally graded material under an anti-plane shear impact load", Key Eng. Mater., vol. 251-252, pp. 123-129, 2003.

[99]   Z. Shao, L. Fan, T. Wang, "Analytical solutions of stresses in functionally graded circular hollow cylinder with finite length", Key Eng. Mater., vol. 261-263, pp. 651-656, 2004.

[100] K. Kokini, S. Rangaraj, "Time-dependent behavior and fracture of functionally graded thermal barrier coatings under thermal shock", Mater. Sci. Forum, vol. 492-493, pp. 379-384, 2005.

[101] H. Yin, L. Sun, G. Paulino, "A Multiscale Framework for Elastic Deformation of Functionally Graded Composites", Mater. Sci. Forum, vol. 492-493, pp. 391-396, 2005.

[102] Z. Zhou, L. Zhang, Q. Shen, D. Gong, "Design model and prediction model on preparation of functionally graded material by co-sedimentation", Mater. Sci. Forum, vol. 492-493, pp. 471-476, 2005.

[103] P. Garrido, R. Burger, F. Concha, "Settling velocities of particulate systems: 11. Comparison of the phenomenological sedimentation–consolidation model with published experimental results", Int. J. Miner. Process., vol. 60, pp. 213–227, 2000.

[104] M. Latsa, M. Assimacopoulos, A. Stamou, N. Markatos, "Two-phase modeling of batch sedimentation", Appl. Math. Modelling, vol. 23, pp. 881-897, 1999.

[105] R. Turian, T. Ma, F. Hsu, D. Sung, "Characterization, settling and rheology of concentrated fine particulate mineral slurries", Powder technology, vol. 93, pp. 219-233, 1997.

[106] R. Bürger, "Phenomenological foundation and mathematical theory of sedimentation–consolidation processes", Chem. Eng. J., vol. 80, pp. 177–188, 2000.

[107] M. Vanni, "Creeping fow over spherical permeable aggregates", Chem. Eng. Sci., vol. 55, pp. 685-698, 2000.

[108] M. Doheim, M. Abu-Ali, S. Mabrouk, "Investigation and modeling of sedimentation of mixed particles", Powder technology, vol. 91, pp. 43-47, 1997.

[109] T. Larsen, "Measuring the variations of the apparent settlingvelocity for fine particles", Wat. Res., vol. 34, pp. 1417-1418, 2000.

[110] Drenchev L., Sobczak J., "Theoretical and Numerical Analysis of Segregation of Solid Particles in Centrifuge Casting Process", J. Mater. Sci. Tech., vol. 7, pp. 34-51, 1999.

[111] J. Richardson, W. Zaki, "Sedimentation and fluidization: part I.", Trans. Int. Chem. Engrs., vol. 32, pp. 78-94, 1954.

[112] D. Shaw, Introduction to Colloid and Surface Chemistry, Butterworth-Heineman, 1999.

[113] L. Drenchev, J. Sobczak, N. Sobczak, W. Sha, S. Malinov, "A comprehensive model of ordered porosity formation", Acta Mater, vol. **55,**, pp. 6459–6471, 2007.

www.ingramcontent.com/pod-product-compliance
Lightning Source LLC
Chambersburg PA
CBHW041723210326
41598CB00007B/759